Principles of Forest Hydrology

Principles of Forest Hydrology

John D. Hewlett

The University of Georgia Press · Athens

ACKNOWLEDGMENTS

This book represents a revision of An Outline of Forest Hydrology (The University of Georgia Press, 1969) written jointly by John D. Hewlett and Wade L. Nutter. The contributions of Dr. Nutter to this revision are acknowledged. A number of graduate students and associates must be thanked for their help in critically reading parts of the manuscript, checking mathematical conversions and continuity: Pierre Bernier, Eulalie Cabanach, Steven Woodall and Ray Doss. Barbara Hawks typed the final copy, and Barbara Daniel designed the cover and drew most of the illustrations.

Athens, Georgia 30602

Printed in the United States of America

Library of Congress Cataloging in Publication Data

Hewlett, John D.
Principles of forest hydrology.

Rev. ed. of: An outline of forest hydrology /
John D. Hewlett and Wade L. Nutter. 1969
Includes index.
1. Hydrology, Forest. I. Title.
GB842.H48 1982 551.48'0915'2 81-16371
ISBN 0-8203-2380-2 AACR2

92 91 90

TABLE OF CONTENTS

PREFACE

The flow of water through the hydrologic cycle establishes the basic level of land productivity. The student pursuing a career in forest and wildland resources soon learns that no natural science is more fundamental to the art of land management than hydrology. A growing body of knowledge about forest and wildland water relations has come to be called forest hydrology.

Hydrology as a science emerged from the various methods used by engineers to predict water flows and to design structures. Many texts focus on technique almost to the exclusion of principle, leaving the effects of land use on water to be interpreted from standard methods rather than from knowledge of hydrological process. The complexity of modern water problems -- shortages, pollution, floods and erosion -- calls for more classroom emphasis on the source area processes that are directly affected by land use, particularly on the soil, atmospheric and vegetal phases of the hydrologic cycle.

This book addresses the natural processes in hydrology. The subject matter is organized in outline form to introduce concepts and principles, and to provide consistent terminology to aid the student in comprehending the current and older literature of hydrology. The text is intended as a base for class lectures and as a guide to further study.

Together with George W. Brown's text on Forestry and Water Quality, published in 1980 by the Oregon State University Book Stores, Inc., this outline should meet the needs of senior level hydrology courses in schools of forest resources, environmental sciences and geography. In addition, state and federal agencies, as well as private industries and foundations, may find the book useful for quick reference in the daily job of policy-making, planning and management of land.

Athens, Georgia
July 1981

John D. Hewlett

CHAPTER ONE

INTRODUCTION TO FOREST HYDROLOGY

A. Purpose. This text will introduce the forest resources student and other students of land management to the principles of forest hydrology, and will suggest how to apply them to management of forests and wildlands for water and water-related resources. Forest resources include all the wood, water, wildlife and amenity values of forests and wildlands, both public and private. The land manager may be interested primarily in forest resources or, on the other hand, he may be concerned with regional planning for all natural resources, only a part of which is associated with forests. A knowledge of the behavior of water in the hydrologic cycle is essential to silviculturists, fish and game managers, range managers, engineers, recreation specialists, land developers and urban planners. The student is advised to approach the study of forest hydrology from the viewpoint of a land or habitat manager, rather than from the narrower perspective of a one-resource specialist.

B. Scope. The content will extend from a review of hydrologic processes and principles to the linking of theory to practice. However, the main subject matter will revolve around the hydrologic cycle, the central concept of hydrology. The hydrosphere embraces the biosphere, and all animate and inanimate things in nature either need water or are strongly affected by its presence, absence or behavior. The sun provides the energy to keep the hydrologic "machine" working, which means the energy cycle also is part of the subject matter of hydrology. The energy in water performs both beneficial and damaging work on its path from cloud to ocean and back again.

C. Definitions. The history of man's study of water and his use of it is as interesting and ancient as any branch of the earth sciences. Man began much of his primitive agriculture as a water manager or water engineer. We still have water managers whose chief purpose is to supply water of sufficient quality and quantity to meet specific uses. The modern water resources specialist, on the other hand, takes a much wider view of the interrelations among man's activities and the hydrologic processes on watersheds. Water managers of the past thought of water as a raw material to be found and transported to the point of use. That was a time of relatively small per capita demand for water, perhaps 20 liters per person per day for all domestic uses. Today per capita demand is estimated at about 700 liters per day, and the complexities of urban-industrial life are expected to push this beyond 1000 liters by the end of the century. It is no longer a simple matter of "find and transport," but a complex land management problem of protection, conservation, control, fair distribution and reuse -- all adding up to multiple-purpose management of forest and wildland watersheds.

Consider the wise words of Voltaire: "If you would converse with me, first define your terms!" All study and understanding begins with good definitions:

Hydrology (the science of water) is the study of water in all its forms (liquid, gas and solid) on, in and over the land areas of the earth, including its distribution, circulation and behavior, its chemical and physical properties, together with the reaction of the environment (including living things) on water itself. The science of hydrology underlies land (watershed) management. The layman sometimes confuses hydrology and hydraulics: Hydraulics is the physics (statics, kinematics and dynamics) of water in its liquid state.

Forest hydrology is a branch of hydrology that deals with the effects of forests and associated wildland vegetation on the water cycle, including the effect on erosion, water quality, and microclimate.

Water management consists of the development, processing, storage and transportation of water for agricultural, industrial, recreational and domestic uses. Examples of water managers are sanitary engineers, irrigation specialists, flood control engineers and reservoir managers. The key concept is of water as a raw material, like coal, wood or stone.

Watershed management is a term used mostly by foresters and soil conservationists and was first defined in "Forest Terminology," Society of American Foresters, 1944:

> The management of the natural resources of a drainage basin primarily for the production and protection of water supplies and water-based resources, including the control of erosion and floods, and the protection of esthetic values associated with water.

Unfortunately this definition does not clearly identify an activity or a manager. The management of natural resources refers to the development and administration of all land resources to satisfy the needs of present and future human residents. In a sense, land, watershed and habitat are synonymous terms; you cannot manage one without simultaneously managing the others. The central point is that resource managers require elementary knowledge of hydrology to be effective in their work.

Forest influences include all effects resulting from the presence of forest or brush upon climate, the water cycle, erosion, floods and soil productivity (Kittredge 1948). Formerly a field in itself (the term comes from Marsh 1863) but now divided between forest hydrology and forest meteorology.

Forest meteorology (forest micrometeorology) deals with the physics of heat, matter and momentum fluxes in the forest biosphere (tops of the crowns to lowest roots). Hydrometeorology refers to the overlap between hydrology and meteorology. Hydrogeology relates to ground-water.

Watershed is any sloping surface that sheds water, defined in Webster's as a topographic divide that sheds water into two or more drainage basins. Despite the "proper" definition, American land managers use watershed as synonymous with drainage basin or catchment. Confusion occurs because the layman often relates to the dictionary meaning, that is, a water divide.

Drainage basin is a watershed that collects and discharges surface streamflow through one outlet or mouth. A catchment is a small drainage basin; a river basin is a large one, but no area limits are specified.

Hydrologic cycle (water cycle) refers summarily to the cycling of water from the ocean to the land and back again, including all the pathways and processes connected with the storage and movement of water in its three states. In a sense, all geochemical cycles are embraced by the water cycle.

Mineral cycle refers summarily to movement of elements and minerals to and from the watershed. The agents are flowing water, wind, gravity and certain biological factors, including man.

Nutrient cycle refers particularly to the movement of plant and animal nutrients, both mineral and organic. These cycles occur between plants, animals and soils, as well as in and out of watersheds, the rate of cycling depending largely on the water economy of the ecosystem.

Water resources are part of natural resources, the goods and amenities afforded man by the environment. Water is used as a chemical, solvent, coolant, source of energy, transportation medium, a recreation base, habitat for fish and wildlife, a silvicultural and agricultural raw material, an agent for waste removal -- in short a material of immeasurable economic and esthetic value. All natural resources might be summed:

Land + Water + Air + Solar energy + Life

Soil and water conservation is a field of human endeavor included in the concept of "watershed management" but specifically devoted to the prevention of soil erosion as it affects water quality and human property, as well as to the preservation of soil fertility and land beauty.

D. The four attributes of water. Throughout this outline, four basic attributes of water are referred to repeatedly: Quantity, quality, timing (regimen) and the energy disposition of water as it cycles through the watershed. Man's activities affect each of these in different ways at different places, sometimes beneficially, sometimes not. It becomes essential to be clear in each case which influence and which attribute is under question. Much confusion has occurred in popular accounts of hydrological influences of forest and land use because of failure to consider specific causes and effects.

E. Historical notes; a brief outline only to guide some additional reading.

1. Earliest efforts to manage water. Water engineering, one of the most ancient of man's cooperative activities, led to the "hydraulic civilizations" in Mesopotamia, Indus and Nile river basins, and in the Negev Desert. Wells, canals, collecting tunnels (kanats), cisterns, laws and dams developed mostly in that order. The Doctrine of Prior Appropriation, a set of rules to allocate water in arid lands, stems from that time: The first to claim water by public notice retains all rights to divert and use water, whether or not his land borders the water course.

2. The science of hydrology was late in coming. It probably began with Aristotle's deduction (about 400 B.C.) that condensation of atmospheric moisture feeds springs and rivers, or that the pressure of the earth mantle on the sea forces water up and purifies it. Though great water engineers, the Romans left little scientific record. The Doctrine of Riparian Rights to control water use in humid lands began with the Romans: Only those owning land to the water's edge have rights to water and these are sold with the land. Leonardo da Vinci (1500) apparently offered the first coherent description of the water cycle, and two Frenchmen, Perrault and Marriotte, were first to measure it on a drainage basin (1680). Halley estimated evaporation rates and used them to compute the water balance of the Mediterranean Sea (1693). Daniel Bernoulli (1738) laid the groundwork for all future advances in the physics of water flows in pipes, streams and artificial channels. George P. Marsh, perhaps father of the conservation movement, devoted a long chapter of his famous book "Man and Nature" (1863) to forest influences, a term he apparently coined, and thus called attention to the hydrological role of forests. The science of hydrology acquired first focus with Daniel Meade's use of the word as title for his text Hydrology (1919). J. Kittredge (Forest Influences 1948, Chap. 2) gives an account of the development of forest

hydrology, including a summary of early notions about forests' effect on rain, water flows and droughts. A. K. Biswas (History of Hydrology 1970) gives a comprehensive account of hydrological history, in which he quotes Aristotle: "He who sees things grow from the beginning will have the best view of them." A modern textbook in general hydrology is that of Ward (1975).

3. Forests, soils, climate and water firmly linked by 1900. Following Marsh's alarming descriptions of the evils of deforestation, there occurred a "propaganda period" of forest influences (1877-1910). The public's perception of these influences, whether real or imagined, played a large role in the forestry and conservation movements. B. E. Fernow ("first American forester"), Hough, Pinchot, Roth and T. Roosevelt, all notable names in forest history, formulated and administered policies which led to many acts of Congress related to forest land and water: The Act to Establish Forest Reserves (1891); Weeks Law (1911), "to protect the headwaters of navigable streams"; McSweeny-McNary Act (1927), authorizing Forest Experiment Stations; New Deal (1932), creating the Civilian Conservation Corps for public works in soil and water conservation; Flood Control Act (1936), giving flood control funds to land management agencies as well as the Corps of Engineers; Soil Conservation Service Act (1935); The Small Watershed Act (1954), forcing involvement of local watershed associations in publicly funded watershed programs.

Water resources problems had become very complex by 1959 when a Senate Select Committee found that water pollution, water resource planning and education were the greatest water problems of the Nation. Since then a flood of acts aimed at planning, research, conservation and development have passed. Important ones were the Water Pollution Control Act (1964 and 1972), the Water Resources Research Act (1965), Water Resources Planning Act (1966), and the National Environmental Protection Act (1972) which created the Environmental Protection Agency and led to many additional acts, agencies and regulations that formed a difficult socio-legal environment for land management during the 1970s, and incidentally called for more hydrological expertise than appeared to be available.

4. A summary of trends in water management in the 20th Century. Developments over the years focus attention on the inescapable relation between forest resources and water resources:

Growing population pressure on land and water.

Unexpected rise in the demand for water created by technology and high living standards.

Increasing knowledge of the hydrologic cycle and man's influence on it.

Emergence of new and complex problems with water; settlement of arid lands, flood plains, pollution, local regional and international conflicts over water rights, conflicts over multiple use of forests and wildlands.

Gradual yielding of property rights to public regulation.

Public awareness of the ecological precept that everything natural and man-made is connected in complex ways.

Managers, planners and developers gradually recognized that the drainage basin is the best natural unit for resource management.

Further readings.

Biswas, A. K. History of Hydrology. North-Holland Publishing Co., Amsterdam, Netherlands, 336 pp., 1970.

Colman, E. A. Vegetation and Watershed Management. Ronald Press, New York, N.Y., 412 pp., 1953.

Fernow, B. E. Forest Influences. U.S. Forestry Division Bulletin No. 7. 1902.

Kittredge, J. Forest Influences. McGraw-Hill Book Co., New York, N.Y., 394 pp., 1948.

Marsh, G. P. Man and Nature. Scribner and Sons, New York, N.Y. 1863. (Reissued in 1902 under title Earth as Modified by Human Action.)

Senate Select Committee on National Water Resources, 86th Congress. Water Resources Activities in the United States. Committee Reprints 1-32, 1959-60. (Particularly Print No. 2, Reviews of National Water Resources During the Past Fifty Years.)

Ward, R. C. Principles of Hydrology. 2nd Ed. McGraw-Hill Book Co., New York, N.Y., 367 pp., 1975.

Zon, R. Forest and Water in the Light of Scientific Investigation. U.S.D.A., Gov't. Printing Office, 106 pp., 1927.

CHAPTER 2

THE WATER AND ENERGY CYCLES

A. The global water cycle. Of the sun's energy that reaches the earth's surface, estimated at about 400 calories per square centimeter per day in the United States, roughly 50% goes to vaporize water. The remaining 200 calories warm the earth, the air and the ocean, producing convectional currents in the atmosphere and the ocean, which in turn distribute heat to moderate climate and to transfer evaporated ocean water to the continents. Thus the sun is the earth's water boy, ceaselessly filling lakes, rivers and soil mantles, and keeping springs and rivers flowing. Water is the earth's largest dynamic heat sink, and if it were not for convectional circulation and the latent heats of fusion and vaporization, our climate would be grim indeed.

Where is the world's water at any one time? According to data assembled during the 1965-75 International Hydrological Decade Program, only about 3% of the total is fresh water, and 76% of that 3% is in the polar ice-caps and mountain glaciers, therefore unavailable for man's ordinary uses. The distribution of water is summarized by Nace (1967):

SALT WATER	%	FRESH WATER	%
Oceans	97.1	Ice and snow	2.2
Salt lakes and seas	.01	Ground and soil	.6
		Fresh water lakes	.01
		Atmosphere	.001
		Rivers	.0001
TOTALS	97.11		2.8111

Two fresh water lakes contain most of the lake water: Lake Baikal in Siberia and Lake Superior in the U.S. Groundwater forms the largest stored volume of fresh water but its use is limited by its distribution, quality and pumping costs. Most withdrawals by man are from the rivers, which, although they hold a tiny amount of the total fresh water, are frequently replenished by condensation of the much larger volume of atmospheric moisture. Over 99% of the world's water seems to be denied to man, but the sun's energy continually renews the 1% remaining. We need have no fear of "running out of water," a popular newspaper phrase that does little to solve water problems, but we need to recognize that constantly rising demands on water must force better planning and pricing of the water resources. That in turn requires better understanding of water.

B. Units essential to water accounting. This outline is written in the metric system but we cannot yet dispense with English units. Four conversion constants must be memorized in order to use manuals, maps, court records, and to converse with laymen:

$1\ mi^2 = 640\ ac$ $\qquad$ $1\ ac = 43{,}560\ ft^2$

$1\ ft^3 = 7.5\ gal$ $\qquad$ $1\ ft^3 = 62.4$ lbs of water

It is best always to compute other units needed. The metric system requires only memorization of the unit names: Kilometer, meter, centimeter, millimeter, micron; cubic meter, liter, milliliter; kilogram, gram, milligram, microgram; and so on. Do not try to memorize conversion constants; look them up in tables each time they are needed (see page 177).

1. The water balance of drainage basins (Eq 2-1) is most often reported in linear measure (cm or inch) expressing gain or loss by precipitation, evaporation or streamflow as a uniform depth of water over the entire basin. Linear expression of streamflow or rainfall imply volumes of water only if the area of the basin is known. Avoid the term "area inch," which merely means "inch." Dimensionally, a vertical length L times horizontal area L^2 equals a volume L^3.

2. Volume of water may be expressed in m^3 or in ft^3, but these are too small for hydrologic use. A hectare-meter (ha-m) is 10,000 m^3, or one level hectare one meter deep in water. The English unit is the acre-foot (ac-ft), one level acre, one foot deep. 1 ha-m = 8.107 ac-ft.

3. Rate of flow (L^3T^{-1}) in streams, soil or pipes is expressed as m^3/minute, ft^3/sec, cm^3/hour, etc. Linear rate is also used (LT^{-1}) as in cm of rainfall per hour or in/hr.

4. Rate per unit of drainage area ($L^3T^{-1}L^{-2}$) is a unit peculiar to hydrology, a way to express streamflow comparative to area. For example, streamflows in eastern U. S. vary from 0.15 to 1500 $m^3/min/km^2$, the former in droughts, the latter in floods. The English unit is $ft^3/sec/mi^2$, which equals 0.656 $m^3/min/km^2$, or 0.011 $m^3/sec/km^2$.

Example. The following diagram represents the annual world water balance in km^3 of water moved from ocean to land and back. (The estimate is from various reports of the International Hydrological Decade, 1965-1975.) From the radius of the earth, the formula for the surface of a sphere and the knowledge that 28% of the earth is land, we know that 146 million km^2 is land, and 375 million km^2 is water. Compute the annual precipitation, evaporation and discharge to and from the land.

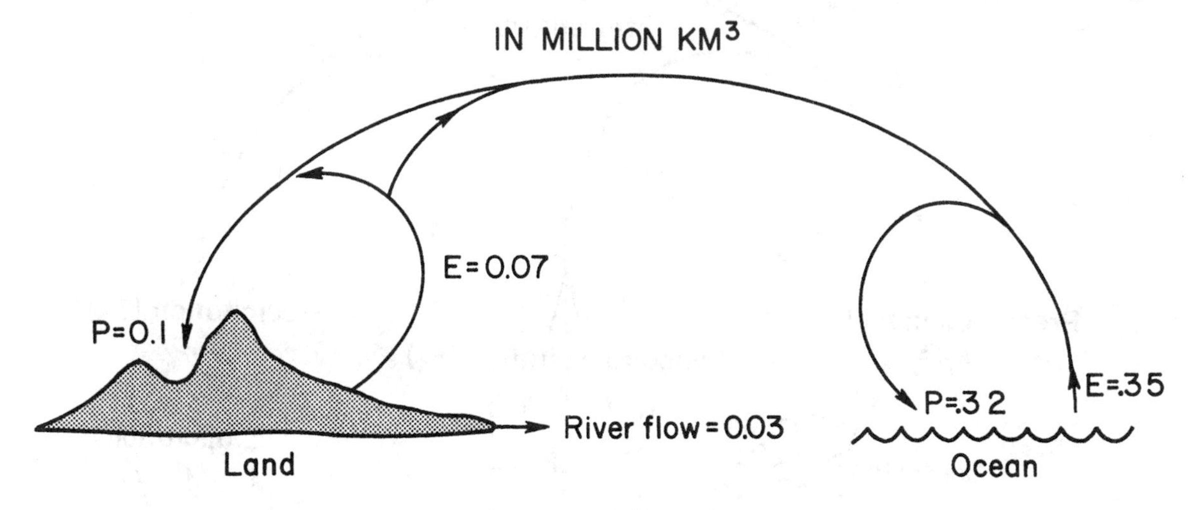

Divide the volumes of water by the corresponding areas:

$$P\ land = \frac{0.10 \times 10^6 km^3}{146 \times 10^6 km^2} (10^3 \frac{m}{km})(10^2 \frac{cm}{m}) = 68.5\ cm$$

$$E\ land = \frac{0.07 \times 10^6 km^3}{146 \times 10^6 km^2} (10^5 \frac{cm}{km}) = 48.0\ cm$$

$$Q\ land = \frac{0.03 \times 10^6 km^3}{146 \times 10^6 km^2} (10^5 \frac{cm}{km}) = 20.5\ cm$$

How many cm fall annually on the ocean?

While it is doubtful that man can seriously alter the world water balance, planetary process has done so. During the build-up of glaciers in the ice ages, the net inland movement of precipitation must have exceeded river flow for thousands of years, lowering ocean levels by 20 meters or more, but the trend reversed as the ice caps retreated, flooding the continental shelves once again.

C. Water balance of the contiguous U.S. Diagrams cannot make the water cycle clear because variations in storages, pathways and rates force us to oversimplify. However, the general water cycle shown in Fig 2-1 suggests circulation pathways. What happens to the average annual 76 cm of precipitation over the 48 states?

$$P_g = E_t + Q + \Delta S \qquad 2\text{-}1$$

$$76\ cm = 53\ cm + 23\ cm + 0 cm$$

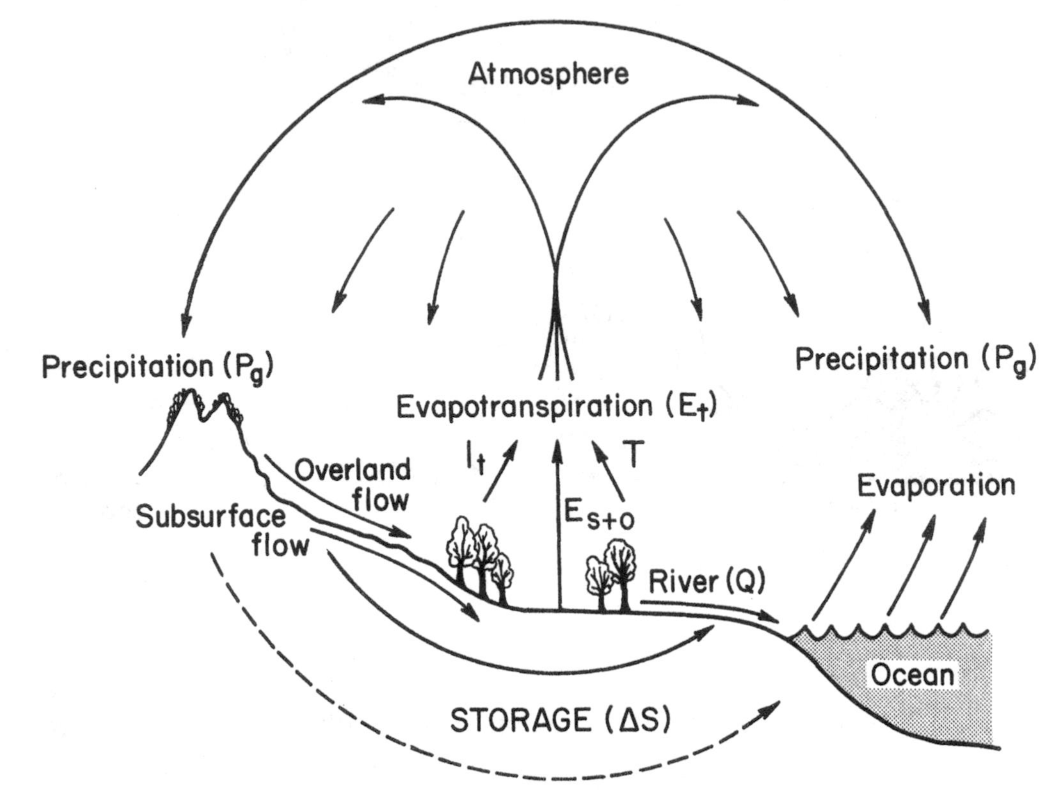

Fig 2-1. The general water cycle. Unnamed symbols are: I_t = interception loss by evaporation; T = transpiration loss by evaporation; E_{s+o} = evaporation from soil and water surfaces.

This is the annual balance estimated by the U.S. Geological Survey from many years of data on precipitation and streamflow. In a long record the average ΔS tends toward zero, but storage varies from year to year. The "use" of water is not as simple an idea as the use of timber, coal, grain or iron ore; water is not "used up" as these are. Several terms are needed to define water usage.

Withdrawal refers to use of water that involves temporary or long-term removal from normal channels or ground water bodies. Most withdrawn water is returned to channels or ground water, but often degraded in quality. Some is evaporated.

Consumptive use refers to use that involves a substantial amount of evaporation in use. Examples are irrigation withdrawals, evaporative cooling and the watering of lawns. Water tied up in plant and animal tissue (particularly cellulose) may be considered consumptive use but the quantity is relatively small, negligible in the hydrologic cycle.

Mined water refers to ground water that is withdrawn at an annual rate which exceeds the average annual recharge by precipitation. Mining water drops the water table year by year, and adds only a tiny fraction to the annual water balance. Though viewed with alarm, ground water mining is common practice in semi-arid areas, where it ultimately leads to severe problems.

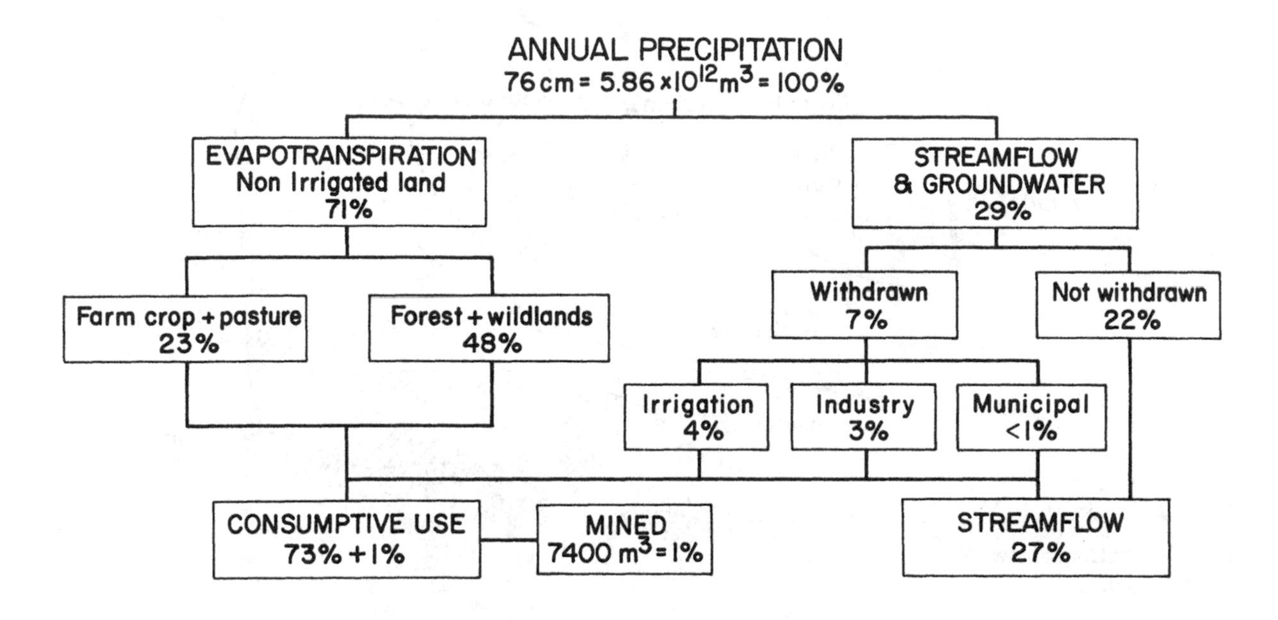

Fig 2-2. Disposition of the annual water supply to the contiguous U.S.

Fig 2-2 shows the annual distribution of U.S. water. Forests and wildlands (tree and brushlands, grasslands, ranges, alpine meadows, deserts, swamps and snow fields) evaporate about half and farmlands (crops, improved pastures, orchards and hayfields) about one quarter of the precipitation before it reaches streams or ground water. The remaining quarter passes through the earth mantle, the rivers and the lakes. About 7% is withdrawn, but all the water supply is "used" in some way, to meet evaporative demands, to fill streams, lakes, ground water bodies, to flush or dilute wastes, to transport goods, to serve habitat and amenity uses.

D. Water balance of drainage basins. Any study of forest hydrology begins with the basin water balance. From the small valley to the great Mississippi-Missouri-Ohio basin we are dealing with discrete land areas that collect precipitation, evaporate water and yield the difference as streamflow or groundwater recharge. Watershed surveys, regional planning and development, hydrologic data collection and experimentation, are best carried out basin by basin. The water balance sets the stage for all effort to improve water supplies, to reduce erosion and flooding, to control nutrient export and pollution, to conserve natural beauty or to develop the energy potential of water. Fig 2-3 shows common terms needed to describe the hydrologic character of the drainage basin, whether large or small.

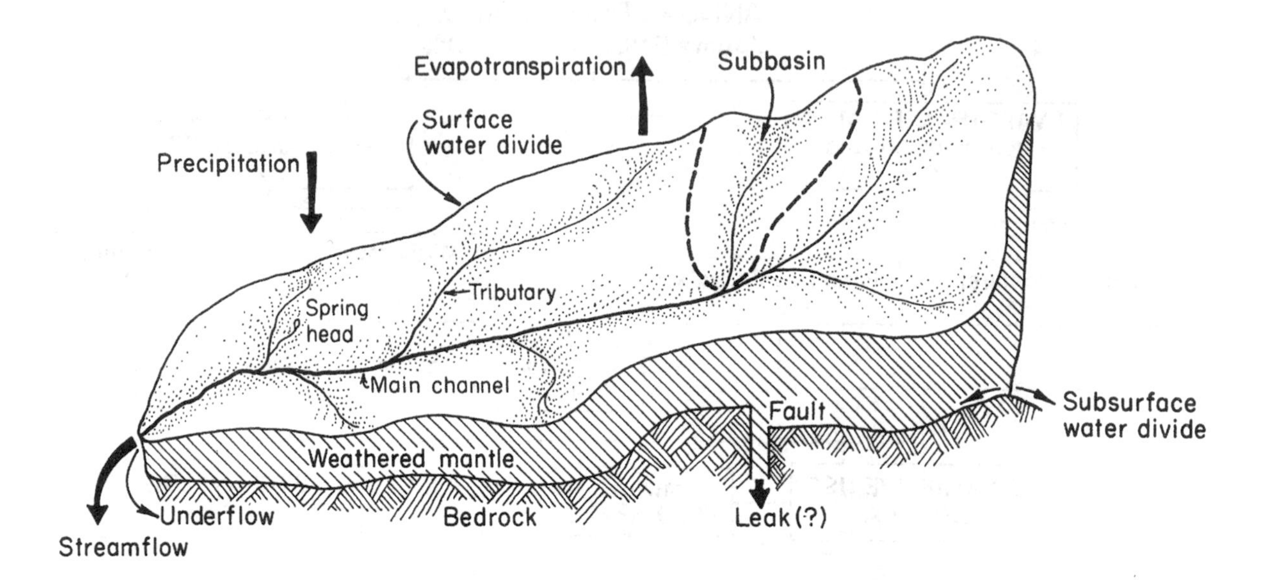

Fig 2-3. Diagram showing terms essential to describe the drainage basin and its water balance. Underflow is unmeasured streamflow in sands, gravels and rocks beneath the channel bed.

1. Terminology in the water balance. Confusion in hydrologic terms have caused problems in communication among land managers. Terms are used to refer simultaneously to rates of flow, volumes of flow, pathways of flow, or simply the process of flow. For example, "runoff" may mean a rate of flow, a volume per unit of rainfall, some water moving over the surface of the ground, or simply the general process of flow within and from a basin. A special effort is made in this outline to provide definitions that will avoid these difficulties. The term runoff, for example, will be used only in the general sense, as in "the runoff process."

E. Energy and water balances related. The principle of the conservation of energy states that all gains and losses of energy must balance at the surface of the earth. One of the largest terms that appears in both the energy and the water balance is the latent heat of vaporization (L). The units of L are calories per gram of water evaporated: 586 calories are required to convert 1 gram of water at 20°C to vapor at the same temperature (and of course vice versa). That is 586 times the calories required to raise 1 gram of water 1°C. Together with the latent heat of fusion, L is among the most important and far-reaching physical phenomena in the biosphere.

1. Three forms of energy enter and leave the watershed:

   Radiant energy, both short wave directly from the sun or reflected from clouds, and long wave emitted by the atmosphere or nearby land masses.

   Thermal energy, both advective in air from adjacent basins and conductive from deep layers of soil and earth. Rain heat adds some energy.

   Kinetic and potential energy of rainfall, the latter representing the mass of water delivered to areas above sea level.

The continuity equation (conservation of energy law) in calories per unit time (or watts per square meter) at the earth's surface is:

$$R_n - H - G - LE_t - p = 0 \qquad 2\text{-}2$$

R_n is net radiation, the sum of all incoming short and long wave radiation from the sun and sky, less reflected short wave radiation and emitted long wave radiation. R_n is usually positive in the daytime and negative at night.

H is the sensible (thermal) heat exchange with the atmosphere; upward flow, usually aided by convection, represents a loss of energy from the earth's surface.

G is the transfer (conduction) of heat through the earth's mass; downward flow is a loss, upward flow a gain to the earth's surface.

LE_t is the latent heat of vaporization, the product of L (586 cal/g at 20°C) and E_t (grams of water evaporated). E_t can be negative if condensation occurs, and the latent heat of fusion (80 cal/g) is added to L if sublimation of ice or snow occurs (Chap 4). Condensation and sublimation are not uncommon on watersheds but the water volumes involved are small.

p represents all the complex chemical energy conversions in photosynthesis (-) and respiration (+). These are hydrologically negligible but biologically fundamental.

Fig 2-4 shows that these components can be positive or negative depending primarily on the time of day. Eq 2-2 assumes that vegetation is part of the earth's surface, which is satisfactory only if viewed from space. In a mature redwood forest the exact location of the surface at which Eq 2-2 is balanced becomes important in explaining the hydrological role of trees. Assume for now that the equation applies at tree tops. Thus one component of G, conduction into the ground, must include conduction into tree trunks. Certain components of evapotranspiration (Chap 6, interception loss and transpiration) are sustained partly by stored thermal heat in soil and vegetal masses (positive values of G in Eq 2-2).

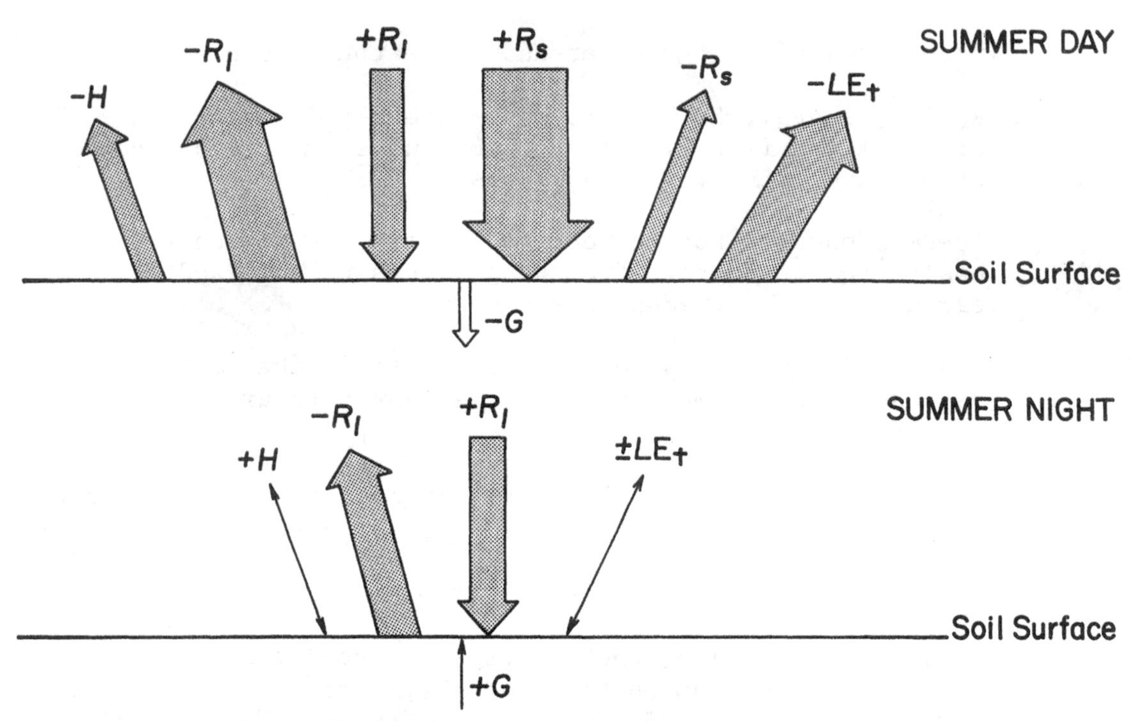

Fig 2-4. Diagram showing the relative energy fluxes at the earth's surface during a summer day and night. Net radiation (R_n) is the sum of the positive and negative components of short-wave (R_s) and long-wave (R_l) radiation.

2. Solving the water-energy balance. If Eq 2-1 and 2-2 are summed over several years for any drainage basin (assuming no leaks or underflow), the change in the storage term approaches zero and may be dropped.

$$E_t = Pg - Q$$

Neglecting the term p, Eq 2-2 may be solved:

$$E_t = (R_n - H - G)/L$$

Set these equations equal to each other and the relatively constant L allows us to compute the balance in calories:

$$(P_g - Q)586 = R_n - H - G \qquad 2\text{-}3$$

Since 1 g of water equals 1 cm^3 of water, we may also read Eq 2-3 in volume (cm^3) of water evaporated from the basin:

$$P_g - Q = (R_n - H - G)/586 \qquad 2\text{-}4$$

Eq 2-4 represents two approaches to studying, explaining and predicting the complex behavior of water and the effect of land use on it. Measurement of the left-hand terms in Eq 2-4 is called the water balance approach, and the right-hand, the energy balance approach, to solving the water budget for land areas.

Further readings.

Lee, R. Forest Micrometeorology. Columbia Univ. Press, N.Y., N.Y., 276 pp., 1978.

Nace, R. L. Are we running out of water? U.S. Geological Survey Circular 586, 7 pp., 1967.

Reifsnyder, W. E. and H. W. Lull. Radiant Energy in Relation to Forests. U.S.D.A. Tech. Bull. 1344, 111 pp., 1965.

Problems.

1. A drainage basin received 48 inches of precipitation during one year. Streamflow was 26 inches and evapotranspiration was 30 inches. Solve for the change in storage in centimeters. Compute signs carefully and show all unit conversions.

2. A ten-hectare field was irrigated evenly with 10 cm of water. How many hours did it take if the irrigation pump operated at 1000 gallons per minute?

3. The average annual precipitation and evapotranspiration on a city watershed are 125 cm and 80 cm, respectively. How large must the basin be (ha) to meet an average demand of 4000 m^3/dy.?

CHAPTER 3

DRAINAGE BASIN MORPHOLOGY

A. Introduction. Geologic structure and the form of the land surface dominate hydrologic processes and strongly influence man's effect on water behavior. Water movement controls present landforms which in turn have been shaped by water in the past. The systematic ordering of the processes that formed land and the study of its present stage of development is called geomorphology. A specialized branch, fluvial morphology, deals with stream and flood plain formation by erosion and sedimentation.

Through study of landforms, their geologic structure and the physical properties of the earth mantle, the hydrologist can explain some hydrologic characteristics and responses, for example, water quality, water storage in the earth, quickness of stormflows and the regimen of streamflow through the year. Knowledge of basin morphology permits recognition of sensitive soil conditions, pollution and erosion hazards, as well as management strategies that tend to reduce costs and increase productivity.

Students of land management are not equal in perception of land form. Designs and prescriptions for erosion control, small bridges, farm ponds, channel improvements and pollution control are often carried out with insufficient knowledge of the structure and operation of drainage basins. Because forest and wildland managers frequently work with basins and subbasins, rather than fields and pastures, they especially need a good sense of drainage pattern -- that is, "the lay of the land." Managerial activities permissible on one part of a basin may not be permissible on another; the efficiency of management in time, money and manpower is directly affected by the spatial pattern of operations.

B. Water and geological control of landforms. Landforms result from the interplay of two opposing forces: uplift and degradation. Uplift is the result of geologic faulting, continental drift, volcanism and wind deposits. Streams and soil-rock movements produce landforms whose shapes depend chiefly on the structure of underlying rocks. The action of water and gravity continue at all times; freezing, thawing, falling, dissolving and depositing. Erosion is a natural process which may be accelerated by man but which cannot be completely halted. Streams cut channels, erode banks and advance up valley slopes by headward erosion.

The transported materials begin a protracted voyage to the sea, leaving erosional patterns behind and producing alluvial plains below. Thus are formed erosional and depositional landscapes, the former frequently in forests or wildlands, the latter more often in agricultural or urban use.

Mass wastage (land slide, rockfall, mudflow, and soil creep), subsequently aggravated by water, is a primary shaper of land in many humid areas, including the Appalachian Mountains and the coast ranges of the West. Subsurface erosion by solution and colloidal transport is often overlooked as an important factor, except in the obvious case of karst topography (very irregular land surfaces formed by collapse of cavities in deep limestones). Estimates vary but assuming an average 20 mg/1 of dissolved minerals in streamwater, the annual river flow of the world (Chap 2) would carry about 4 metric tons per square kilometer (40 kg/ha) to the sea each year, by no means a negligible quantity in geologic time. Subsurface erosion determines natural water quality and nutrient cycling more than any other factor. For example, streams draining limestones are unsuitable for evaporative cooling and steam boilers, while they serve as excellent habitat for a great variety of aquatic organisms.

Important hydrologic features controlled by geological process and structure are the length, angle and depth of weathered mantles on basin slopes, as well as the physical properties of the mantle that largely determine quantity, quality and timing of water yields. A geological name for the weathered mantle is regolith, literally meaning "mantle rock." The depth and physical properties of the regolith constitute the storage capacity of a basin. Of all the hydrologic features of a basin, storage capacity is the hardest to measure but the most interesting to the hydrologist (Chap 7).

C. Stream networks and their classification. In running from the land to the sea, water organizes itself into drainage networks which reflect original geological structures and are bounded by topographic or surface water divides to form basins. Over geologic time, channel networks evolved so as to dispose of water's potential and kinetic energy in the most efficient manner. Leopold et al. (1964) concluded that stream channels tend to take the profile that disposes of the water's energy equally along their length. If there is a sudden drop in the channel, as at a water fall or the head of a gully, the water's energy chews away at the channel discontinuity in an effort to reduce it to a smoothly declining profile. Niagara Falls dramatically exhibits this process by periodic collapse and gradual retreat upstream. Rock formations that resist erosion eventually form nickpoints (rock sills, riffles and cataracts); nickpoints along river networks set basic grades for plateaus, valleys and inland swamps. In arid and semi-arid regions the structural control of landforms is fairly obvious because weathering is not deep and vegetation does not mask rock forms. The relative resistance of rocks determines the drainage pattern and arrangements of slopes and valleys, but differential weathering, mass wastage, faulting and volcanic deposits often conceal the actual bedrock configuration beneath the surface. Therefore drainage basins, particularly small ones, are sometimes difficult to delineate on maps or the ground. Stream networks are easier.

1. Duration of streamflows. All classifications of streams depend on what we call a stream. For example, if streams were of constant length and clearly mapped there would be less disagreement among land managers about stream protection zones. But streams vary in length from season to season and from storm to storm.

> A stream, live stream or perennial stream carries water almost year-round (90% of the time or more) in a well defined channel.
>
> A wetweather or intermittent stream generally flows only during the wet season, perhaps a few months per year.
>
> A dry wash or ephemeral stream flows during and for short periods following rain or snowmelt; channels are often not well defined. Draw, gully, swale, arroyo and gulch mean about the same as dry wash.

Streams should be classified and mapped before land management prescriptions are written.

2. Drainage patterns. Looked at from space most stream channel networks exhibit a tree-like, or dendritic, pattern. But viewed closer, as for example on U.S. Geological Survey 7 1/2-minute topo sheets, the pattern may appear dendritic, rectangular, trellis or radial. Dendritic networks occur in uplands where the regolith and bedrock offer relatively uniform resistance to erosion. Tributaries branch and erode uphill in a random fashion, resulting in slopes with no dominant orientation. Rectangular patterns occur in faulted areas where streams follow the fault lines. Trellis patterns occur under two conditions: One where rocks of unequal resistance are arranged in ridges or folds, as in the Valley and Ridge Province of eastern U.S. and another in broad gently sloping plains of relatively uniform resistance, as in the Southern Coastal Plains. Radial patterns are associated with free-standing mountains such as volcanoes. These terms are useful primarily in describing the texture of the landscape.

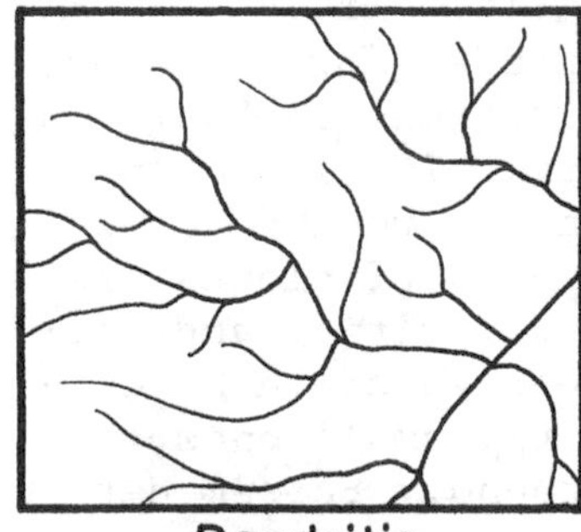

Dendritic

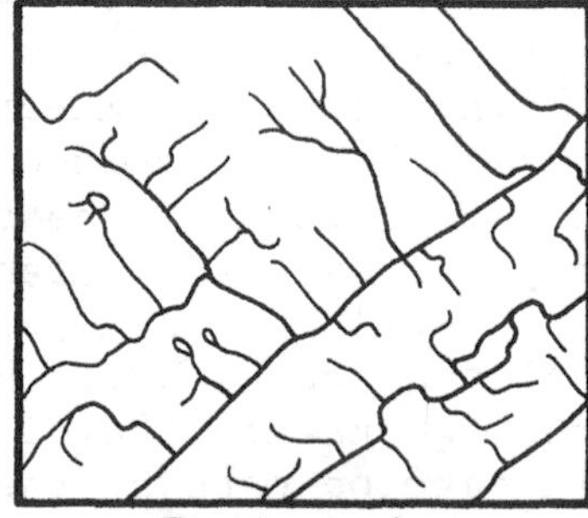

Rectangular

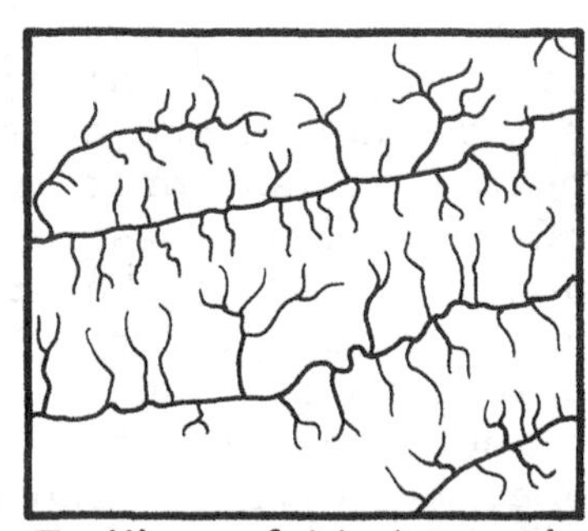

Trellis on folded terrain

3. Stream orders. Stream size may be classified by systematically ordering the network of stream branches. We may begin ordering with either ephemeral, wetweather or perennial streams, but cannot switch back and forth. As shown in Fig 3-1 each nonbranching channel segment is designated a first order stream. Streams receiving only first order reaches are termed second order, and so on. Basins may also be referred to as first order, second order, etc. In the Appalachian Mountains, first order perennial streams have basins from 15 to 25 hectares, while in the Coastal Plain they may be as large as 3 or 4 km^2. In the arid west, first order perennial streams may drain hundreds of km^2. In arid areas, ephemeral channels usually serve as first order.

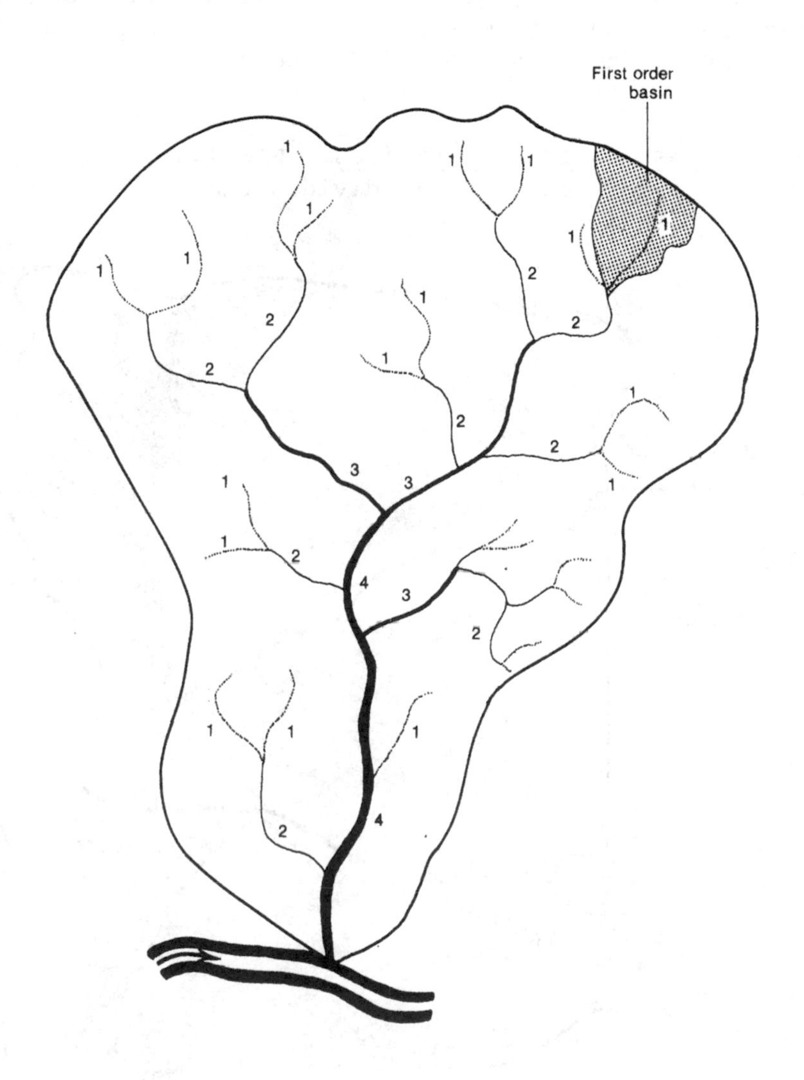

Fig 3-1. Stream orders according to Horton's system of classification.

4. Drainage density. If based on perennial streams, drainage density is an index of how well watered the land is; if based on ephemeral channels, it may indicate how "flashy" the storm response will be.

$$\text{Drainage density} = \frac{\text{Length of channel}}{\text{Area of the land}}, \text{ or } \frac{\text{km}}{\text{km}^2}$$

Drainage density is not a constant for a basin but grows and shrinks during storms and droughts, respectively. However, it is easy to measure on topo sheets, which assume constant stream lengths. The index reflects the average distance to the nearest water, and so has value in assessing wildlife habitat, in laying out roads and stream protection zones, and in rating fire hazards. At drainage density 10, we can determine by trial and error that the average distance to water is about 25 meters, whereas at 1.0, the average distance is about 250 meters.

D. Surface and subsurface water divides. Fig 2-3 (p 12) shows surface and subsurface water divides but not all divides can be so ideally depicted.

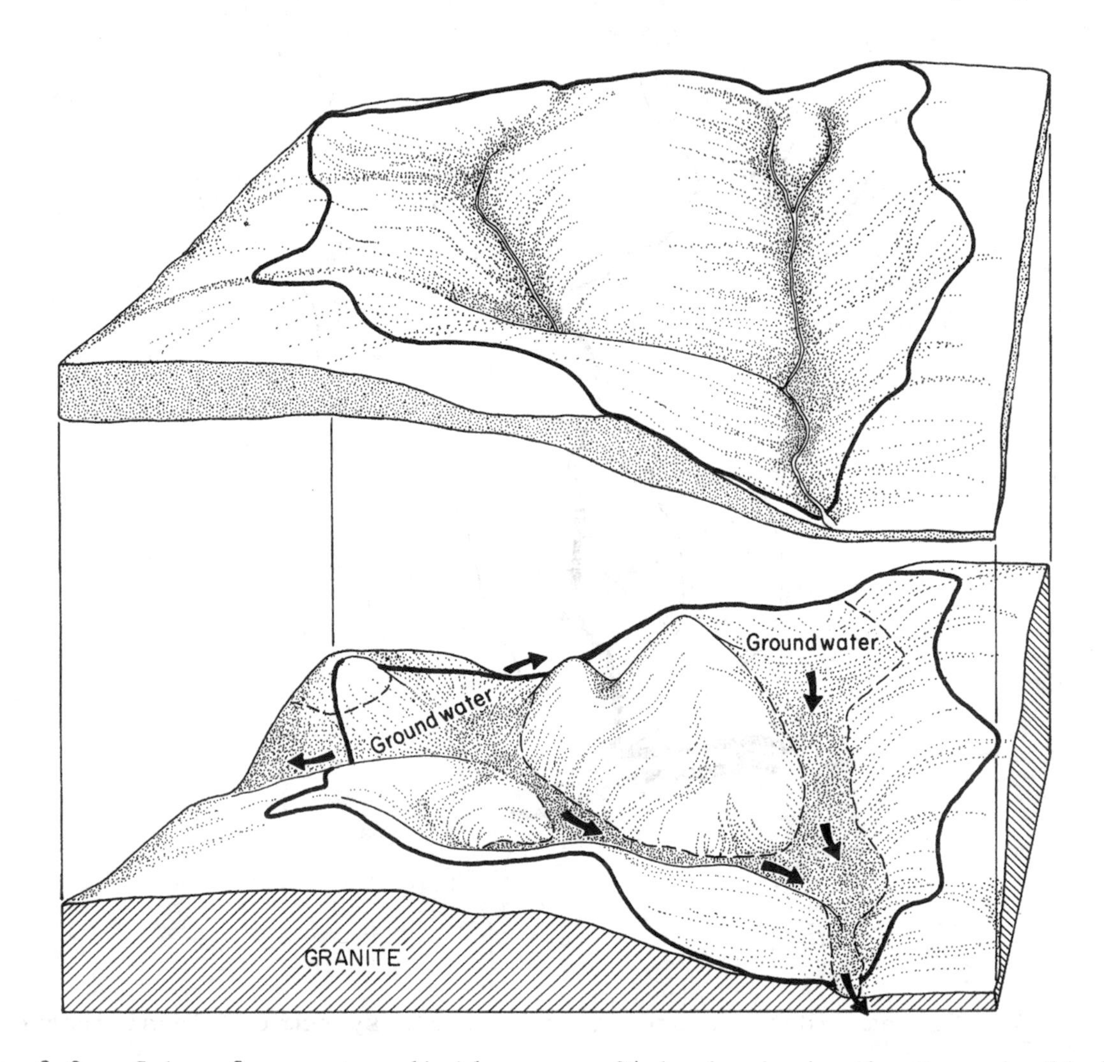

Fig 3-2. Subsurface water divides on a 24-ha basin in the Georgia Piedmont.

In many geologic structures, the two divides do not coincide and therefore water balance accounting is difficult because we cannot satisfactorily determine the area contributing to baseflow from underground sources. Fig 3-2, representing actual measurements made on a 24-ha basin in the Georgia Piedmont, shows the surface water divide projected onto the granitic bedrock below. The problem is greater in small basins than in large ones because the percentage error in estimating subsurface source areas can be much larger in small basins.

In regions with interbedded, tilted or very flat strata of differing permeabilities, subsurface water divides rarely coincide with surface ones. Examples are folded limestones, glacial morraines, lava beds and coastal wetlands. Percolating water may move out of one basin, possibly appearing in the next.

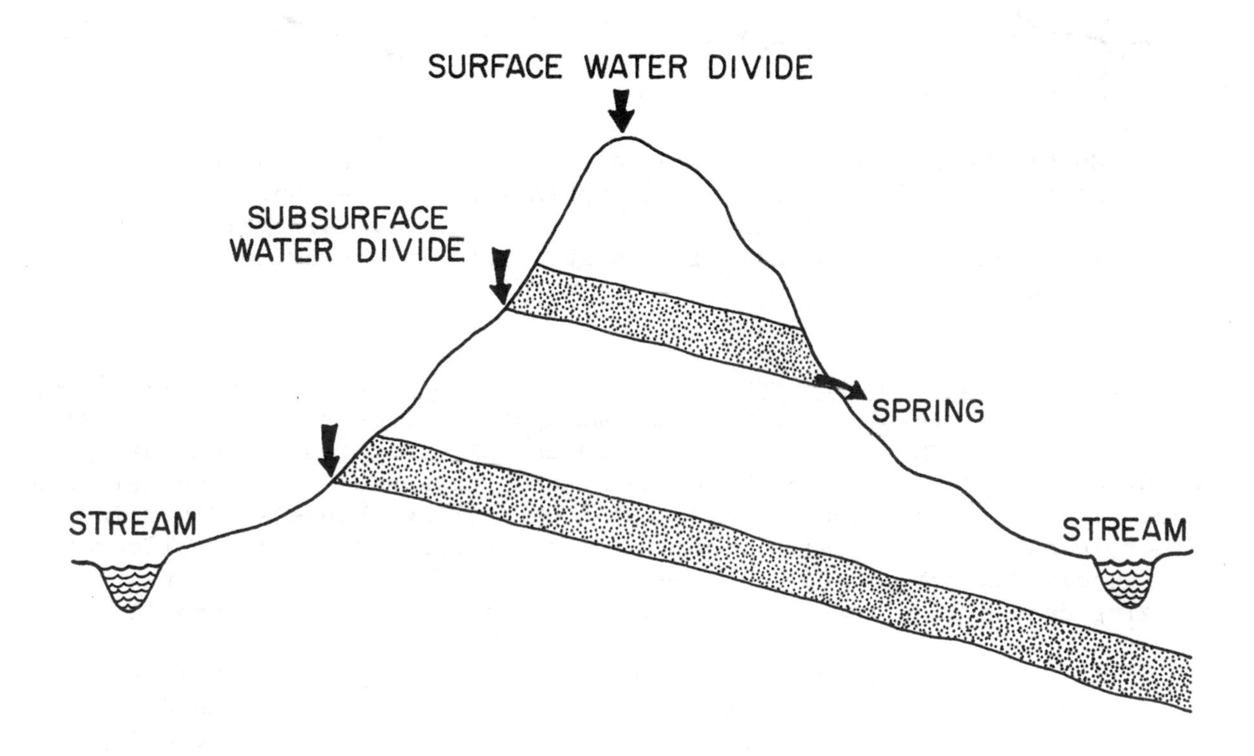

The stippled stratum is relatively pervious compared to the remainder of the earth mantle and water is "leaked" out of the basin on the left. It should be apparent that the larger the basin, the more chance that the inward and outward leaks will balance. In coastal flatlands of uniform permeability, subsurface water divides may wander due to lowering water tables in one direction, or unusually heavy rainfall in another.

1. Hillslopes viewed in profile run from the water divide to the stream channel. Their shapes may be convex, concave, straight line, or a combination of these. Some examples are:

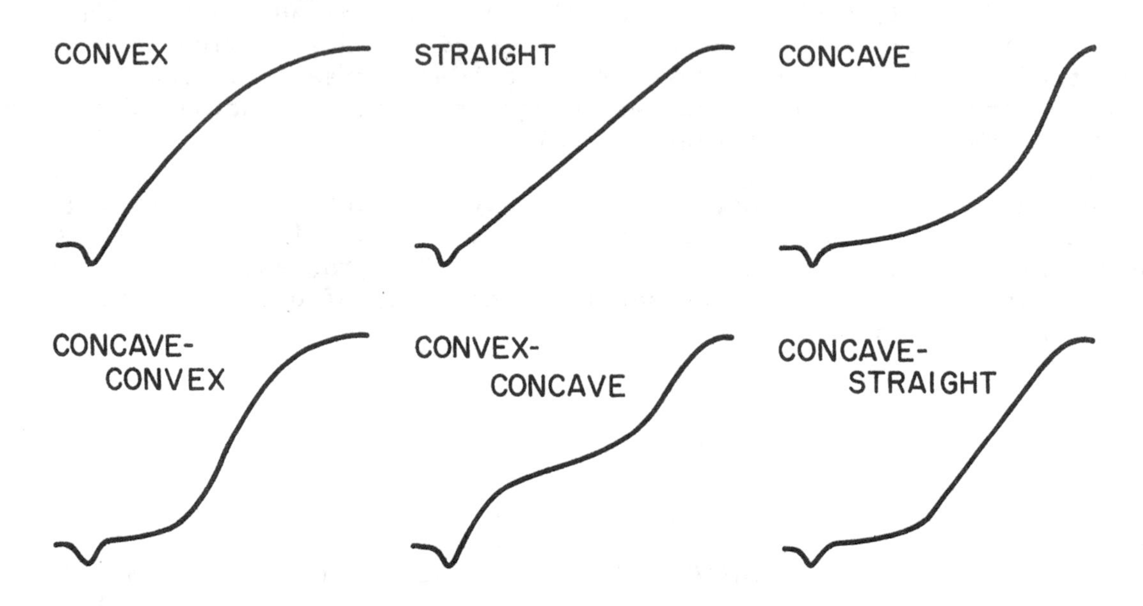

Mostly because of the influence on water behavior, these shapes affect soil depth, site class, optimum road locations, harvesting methods and various other land use activities. A particular basin may exhibit several hillslope shapes from its headwaters to its mouth.

E. Stream channel morphology. Fluvial morphology (Leopold et al., 1964) is in itself a field of study. Land managers dealing with large streams (4th order or larger) must be more than casually concerned with the hydraulics of streamflow and the channel forms that are evolved and altered over time. Lack of caution in dredging or filling channels, diverting flows, constructing erosion control works, altering pool and riffle ratios in fishery management, clearing channel banks of vegetation, or piling logging debris in streams often lead to unexpected results, even lawsuits.

1. Natural levees are a feature of large stream banks that tend over time to destabilize channels and flood plains. Caused mainly by the drag of bank vegetation on flood waters, levees form in a ridge of deposited sand on each side of the thalweg, or central pathway of flow:

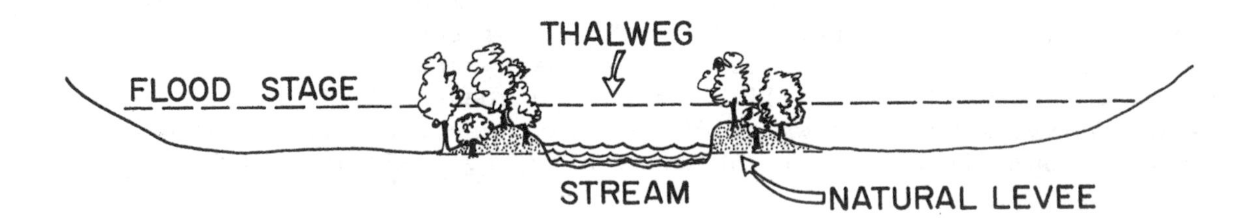

The vegetal drag slows the water along the banks and causes some of the suspended sand to drop out, while the water velocity in the thalweg is sufficient to keep the sediment moving. Eventually, however, a meander breaks through the levee and cuts a new channel in the flood plain, possibly destroying valuable bottomlands, and certainly causing an accelerated cycle of erosion and sedimentation.

2. Stream meanders. Large streams exhibit snake-like curves that are said to characterize old age, or late stages in the uplift and degradation cycle. The causes of meandering are complex, involving transfers between the potential and kinetic energy of water. In a general sense, meanders seem to be a consequence of the stream's effort to dispose of its energy equally along the channel as it lowers itself to sea level. As velocity builds in a straight reach of stream, water resists a change in flow direction and thus chews relentlessly at any bend in the banks. Eventually this pattern forms:

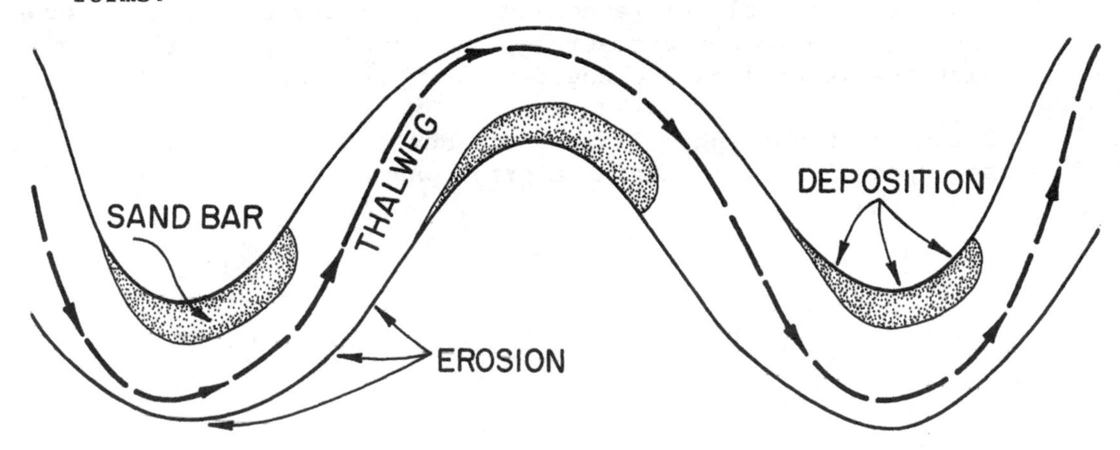

Unless held by a nickpoint or other obstruction, the meanders do not stay in the same place, but migrate slowly downstream reworking the ancient sediments of the flood plain and shifting from side to side. These processes can be resisted by engineering works for decades at a time, but fundamental forces will have their way in the end. Taming stream channels nearly always requires heavier engineering than the resource manager at first thinks.

Further readings.

Heede, B. H. Stream dynamics: An overview for land managers. 26 pp., U.S.D.A. Forest Service, Gen. Tech. Rep. RM-72, Rky. Mtn. For. and Range Exp. Sta., Ft. Collins, Colo. 1980.

Leopold, L. B., M. Gordon and J. P. Miller. Fluvial Processes in Geomorphology. W. H. Freeman and Co., San Francisco, Calif. 522 pp., 1964.

Strahler, A. N. Physical Geography. 3rd Ed. John Wiley, New York, N.Y., 534 pp., 1969 (Chaps. 18 and 24 through 29).

Problems.

1. Secure a 7 1/2-min quandrangle topographic map characteristic of your locality. Perform the following exercises.

    a. Assume that the "blue lines" represent 1st order streams. Determine the highest order stream that occurs on the map.

    b. Select a point on a first or second order stream and trace in the surface water divides above that point. Determine the area of the basin in acres or hectares (use simple dot grid).

    c. Determine the drainage density in miles per square mile.

    d. Using a dot grid, determine the average distance from random points to the nearest stream water. Select the coordinates of a dot from a table of random numbers, locate the dot, measure the distance to nearest stream, and continue until the mean distance is no longer changed greatly by additional observations.

    e. Using the topographic lines and a scale, plot the profile of a selected hill or mountain on graph paper. Classify the hill-slope shape.

CHAPTER 4

ATMOSPHERIC MOISTURE AND PRECIPITATION

A. Atmospheric moisture. When precipitation reaches the biosphere, it becomes the basic element in hydrologic process. However, rain, snow, sleet and hail are not the only forms of atmospheric moisture important in hydrology at the earth's surface. The sum of all vapor, mist and ice in the atmosphere at any one time, representing about 25 mm of water if condensed at the earth's surface, is 10 times that in the rivers of the world (Chap 2). From this transient supply comes all precipitation.

But atmospheric moisture has other roles as well. It moderates temperature extremes because it absorbs or reflects about half of the incoming short wave radiation during daytime, and helps trap outgoing long wave radiation night and day. Moisture in the air also dampens evaporation at the earth's surface, thereby favoring soil water storage and streamflow.

1. Atmospheric humidity. As air becomes warmer its capacity to hold water vapor increases. On the other hand, as air cools cloud particles may grow, coalesce and fall as rain or snow. These thermodynamic properties underlie all terms describing water in the air.

Saturation vapor content is the maximum quantity of water vapor that can be held in a parcel of air at a given temperature and ambient pressure, expressed either as maximum specific or absolute humidity.

Specific humidity is the ratio of water vapor mass to the total mass of the air containing it (g/kg).

Absolute humidity is the ratio of water vapor mass to the total volume of the air containing it (g/m^3).

Specific humidity of a given air mass does not change with changing pressure, whereas absolute humidity does. The density of dry air at sea level is about 1.276 g/m^3. The absolute humidity at sea level is usually less than 0.005 g/m^3; that is, the atmosphere is less than 1/2% water.

Vapor pressure. The pressure exerted by each gas in moist air is the partial pressure. The partial pressure for water is called vapor pressure (e). Pressures are expressed in bars (b) or millibars (mb):

1 b = 1000 mb = 0.987 times atmospheric (barometric) pressure at sea level

1 b = 10^6 dynes/cm^2 UNITS: $ML^{-1}T^{-2}$

Saturation vapor pressure (e_s) is the partial pressure of water vapor at the saturation vapor content. In the normal temperature range, e_s is a known quantity:

Air Temperature C°	Saturation Vapor Pressure mb
10	9.21
20	17.54
30	31.82

Here we see clearly the increased capacity of warm air to hold water vapor.

There are two ways to express the current vapor pressure of the air (e) relative to the saturation vapor pressure (e_s):

Relative humidity (RH) is a ratio of pressures, or (e/e_s) x 100. (A common error is to assume that RH is the amount of H_2O in air.)

Vapor pressure deficit is the difference (e_s - e) in mb.

Dew point (frost point) temperature occurs when moist air below the saturation vapor content at a given temperature is cooled until the saturation vapor content of a lower temperature is reached. Further cooling (below the dew or frost point) results in condensation, causing dew or frost to form on surfaces, or raindrops or snowflakes to grow and fall to earth. Under some circumstances air can be several degrees cooler than the dew point without condensation (supercooled air), but only temporarily.

Adiabatic lapse rate. Together with the adiabatic lapse rate, these static principles and definitions account for the processes of rain, snow, hail, dew, frost, evaporation and atmospheric circulation. "Lapse rate" refers to the change in air temperature with elevation above sea level. Under average conditions the lapse is about 6.5°C per 1000 m, but conditions are seldom average. If a parcel (say 1 kg) of air at 30% RH, containing 5 g of vapor, rises from sea level without gaining or losing water, its temperature will drop "adiabatically." (Adiabatic means without losing or gaining energy; the temperature falls because the energy of the rising air goes into molecular activity of the expanding gases, not into maintaining the temperature of the air parcel.) From the definitions above you can see that the RH will rise as the parcel rises. Finally, the expanding parcel will reach the dew point and cloud droplets will form. Further rise produces rain or snow, sleet or hail depending on how unstable the atmosphere becomes.

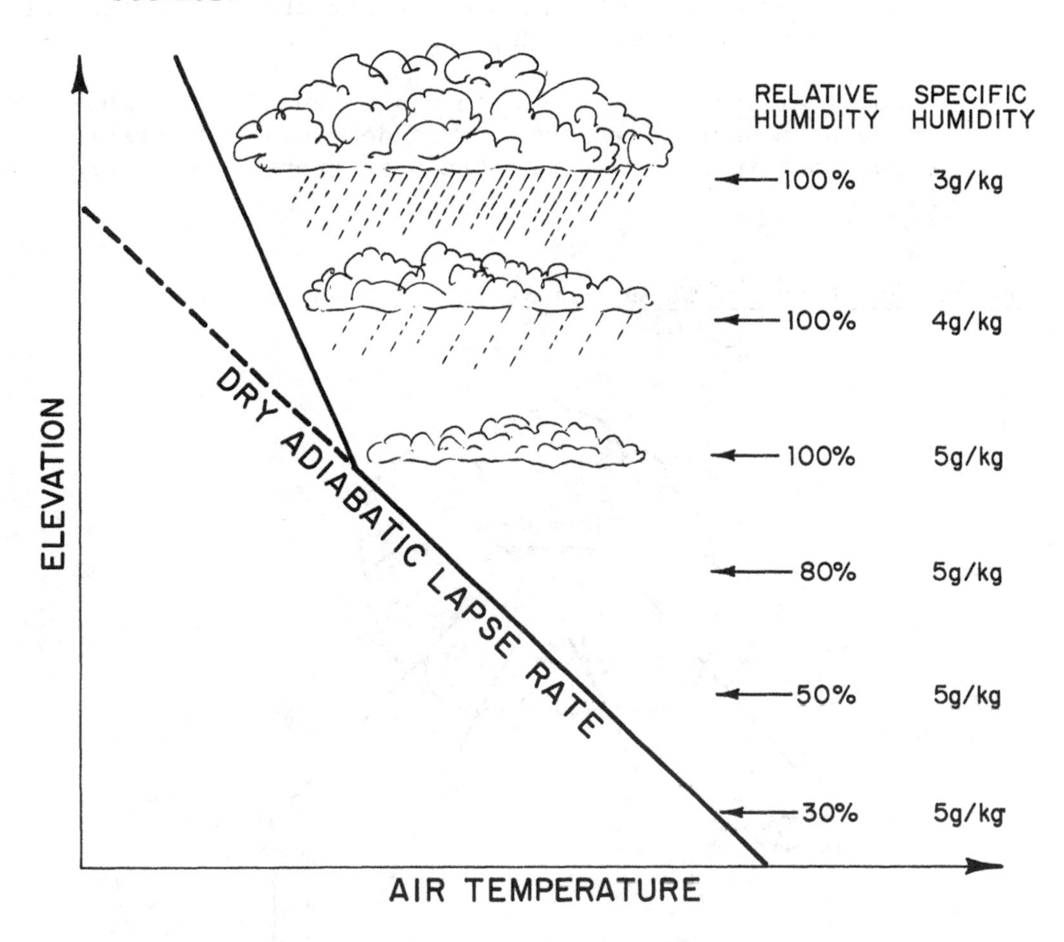

At 100% RH the parcel's temperature lapse departs the dry adiabatic rate because the latent heat of condensation (same calories/g as the latent heat of vaporization) slows the rate of lapse. Thus the rain-making process feeds on its own energy of condensation, pushing the parcel higher and higher until some of its water is squeezed out. A number of weather conditions can be explained by this process, among them convectional storms, frontal storms, hail storms, chinook winds and the like.

2. Devices to measure atmospheric humidity. Vapor cannot be seen but can be sensed by psychrometers and hygrometers.

Wet bulb depression measures relative humidity with a psychrometer, the latter a pair of thermometers mounted side by side, one dry and one kept wet while a thin layer of pure water evaporates from it. The dry bulb temperature and the depression of the wet bulb reading below it are used to compute RH from psychrometric tables.

Hygrometers employ animal skin or human hair to measure RH. These materials, which expand or shrink as they absorb or lose water vapor, are mounted so as to activate mechanical indicators. Hygrometers are calibrated with psychrometers. A hygrothermograph is a common weather station instrument that records both RH and dry-bulb temperature.

Reading electronic psychrometers are needed if air humidity is to be measured hour by hour to determine evaporation conditions in the crowns of trees or crops. There are many kinds.

B. The energy status of water can be illustrated in a diagram.

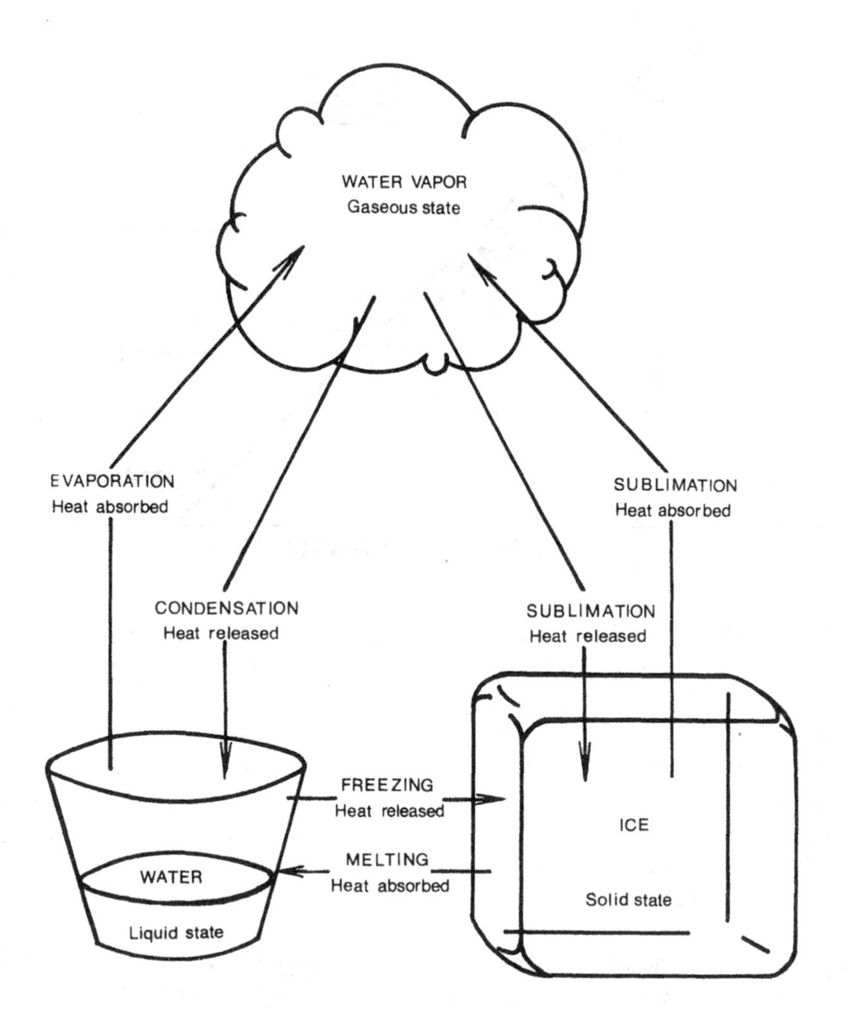

Specific heat of water is 1 cal/g/°C, that is, it requires one calorie to raise one gram of water one degree centigrade, and vice versa.

Latent heat of vaporization and condensation (about 586 cal/g) can take place at any temperature, and is a function of temperature:

$$L = 597.3 - 0.56\ (^\circ C) \qquad 4\text{-}1$$

Latent heat of fusion, 80 cal/g, since it occurs only at 0°C, is unaffected by temperature.

Latent heat of sublimation is about 597 + 80, or 677 cal/g of ice turned into vapor, or vice versa. Sublimation can take place at any temperature but the rate per gram is not strongly affected by temperature.

C. Precipitation occurs when large masses of moist air have steadily moved aloft either due to convectional rise, or to invading warm or cold air masses that slide over or under one another, or due to forced rise over land obstructions. Night-time cooling due to radiant heat loss to space affects surface temperatures enough to cause dew or frost, but does not cool air masses enough to cause significant rain or snow.

1. Types of storms. The three general mechanisms causing uplift of air masses serve to classify storm types:

Convectional storms are caused by differential solar heating of the ground and lower air layers, occurring typically on late summer afternoons when warm moist air covers a region. Convectional cells often cover only 25 or 30 km^2, but rain intensity is high because condensation in the rising moist air pushes it to elevations of 15000 m or more. Hail often forms at these elevations. Convectional cells may occur within frontal storms as well.

Frontal or cyclonic storms are usually caused by warm air invading cold air, or cold air invading warm. In either case the warm air is forced up, producing longer and heavier precipitation over a wider area. Hurricanes, typhoons, and monsoons are of this type, producing seasons of heavy recharge to ground water, reservoirs, lakes and rivers.

Orographic storms are caused by moist air riding over a topographic barrier, such as a coastal mountain range. Distribution of precipitation is strongly affected. In the southern Appalachian Mountains, orographic effects cause a rise in average annual precipitation of about 8 cm per 100 m rise in elevation (about 10 inches per 1000 feet). Orographic effects redistribute the water delivered by a frontal storm.

2. Forms of precipitation. The greatest difference in precipitation form is between rain and snow. The hydrology of regions may be characterized as rain-dominated or snow-dominated.

Rain is formed when condensation droplets become too large to remain suspended. The falling drops grow by collision and condensation up to about 6 mm in diameter before breaking into smaller droplets under the force of acceleration.

Snow forms from water vapor that may be cooled slightly below the frost point, usually near 0°C. In high mountains and northern latitudes, much of the annual precipitation occurs as snow, which piles up to form snowpacks.

Drizzle is a form of rain with droplets less than about 0.5 mm in diameter, characteristic of cool weather and maritime climates.

Sleet consists of small droplets of frozen rain, first formed in upper warm air, but falling through a lower layer below 0°C.

Hail occurs as balls of ice formed by up and down drafts in the higher layers of convectional storms. While suspended in updrafts, concentric layers of ice build up, not infrequently to the size of golf balls. Hail can totally defoliate forests and ruin crops.

Glaze (ice storms) is a coating of ice on the ground or vegetation, formed as rain strikes a surface covered by air that is below 0°C. Glaze sometimes destroys pine plantations.

Rime is a striking display of sublimation that occurs when water vapor is supercooled and blown against twigs, wires, etc. The sublimed vapor builds up on twigs toward the wind as much as 10 cm, frequently turning the tops of mountains into glistening crystal when no other form of precipitation has occurred. The amount of water involved is negligible in hydrology.

Mist precipitation (fog drip) occurs when the wind blows fog or mist through vegetation and the foliage, twigs and trunks intercept the tiny droplets. The droplets coalesce into fog drip or run down the stems to the ground, representing an input to the watershed that is not measured in precipitation gages. In some narrow coastal regions and high mountain fog belts, fog drip is thought to add substantially to annual precipitation. However, measurements are few and the subject of mist precipitation is colored by propaganda that favors the role of trees in producing more water than they consumptively use.

Dew and frost form in small amounts when surfaces are radiantly cooled below the dew point. Negligible in the hydrologic cycle, dew may aid plant survival in semi-arid areas, where much has been claimed for its beneficial effects. It is thought that dew and frost form from water vapor that has only recently evaporated from the soil near the condensate; if so, these are not true forms of precipitation (additions to the watershed).

D. Precipitation measurement. The total amount of precipitation (by storm and by time period), the hourly rate of delivery (intensity), the duration and the distribution of all these over space are subject to measurement. These measurements are highly useful in planning and design, but accuracy and length of record always leave much to be desired. Daily, monthly and annual amounts are published by the Weather Service for about 13000 gaging stations in the U.S., amounts reported as the depth of water accumulated in standard 8-inch (20.32 cm) diameter raingages. The ideal raingage network might provide one gage for every 25 km^2 (size of average convectional cells), which would require almost 3 million gaging sites in the U.S., or 230 times the number now gaged. It is simple to measure rainfall but difficult and expensive to maintain a gaging station. Snow measurement is especially hard.

1. Rain gages. Any water-tight container may be used to measure rain but there is variation in the catch among unlike gages at the same site. The standard 8-inch orifice gage in the U.S. is depicted in Fig 4-1.

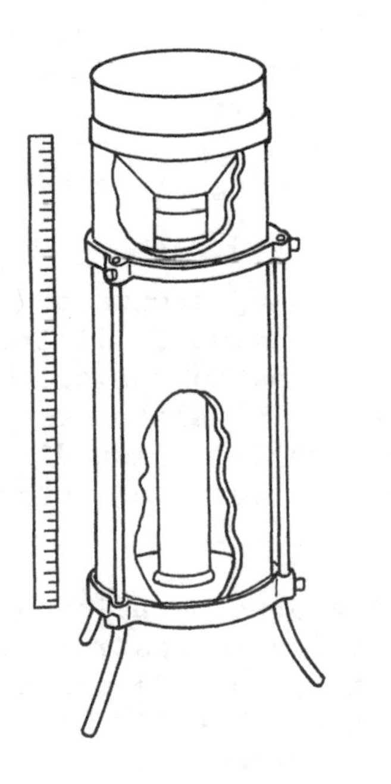

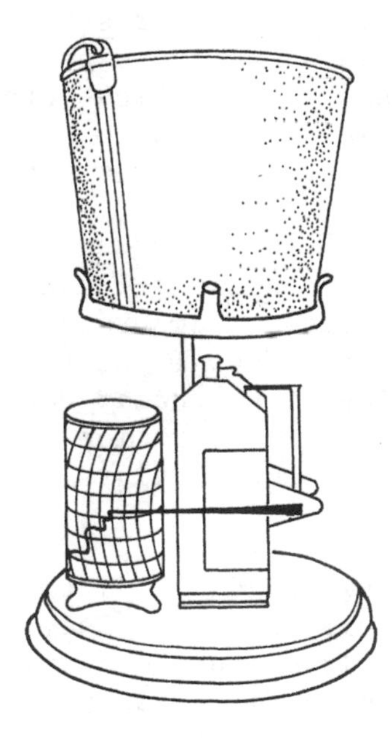

Fig 4-1. Left: Cut-away of the standard Weather Service raingage. Right: A weighing-type recording gage with its cover removed to show the spring housing, recording pen and storage bucket.

Recording gages operate on either weighing or counting principles. The weighing-bucket type (Fig 4-1) records the weight of water on a clock-driven drum, usually limited to 8 days. A tipping-bucket gage records each time a small cup fills and dumps by tipping back and forth. A siphon gage records the filling of a tube which then drains itself by siphoning the water out, thus starting over. Any of these types may be remotely recorded electrically or by radio, the record traced by analog or punched on computer-compatible tape. The weighing-bucket gage is preferable in regions of high rainfall intensities (12 cm/hr or greater) because the tipping bucket and siphon cannot normally keep up with such high intensities.

Non-recording gages may be of any size as long as the orifice is sharp-edged and stable in dimension. Experiments have shown that gage-catch is not very sensitive to size of opening but that the aerodynamics of the opening is a serious factor. A standard 8-inch gage is usually used to correct an adjacent recording gage because the recording devices are often temperature sensitive or in poor adjustment.

2. Snow gages. Because snow is blown about so easily, snow catch in a raingage is a poor sample of snowfall. In regions of low snowfall, recording and standard gages are used after slight modification. The snow is melted and measured in depth of liquid water. In high mountains and northern latitudes (snowpack regions), special storage or standpipe gages are installed so that the orifice is above the expected maximum snow depth, 5 m or more in places. The orifices are shielded in several ways to reduce wind turbulence, but snow still piles up and falls out of the gage or blows straight across it. Errors in snow gaging are often 50% or more.

3. Radar sensing is in common use for tracking storms and violent weather but its use for precipitation gaging is limited to extending actual point records of rain or snow on the ground. Density of the radar pattern is not accurately related to intensity of precipitation because such things as drop size distribution, ground barriers and wind turbulence affect the pattern density.

There are two fundamental problems in measuring precipitation which have not been fully solved. They are (1) how to design an instrument or gage to accurately measure precipitation at one point, and (2) how to locate a network of gages to sample an area within acceptable limits of error. The first may be called instrumental error and the second, sampling error. The design of an error-free gage is not easy. The main problem is wind turbulence around the orifice but splash-in or splash-out is also serious. Sampling error involves the location of gages to sample precipitation on a basin, including errors due to barriers (a hill, a wall of trees, for example), orographic effects and the peculiar wind patterns in small forest openings.

Precipitation measurement in and around forests presents special problems. Gages are usually located in openings cut large enough to prevent tree crowns from projecting into a 45-degree line-of-sight from the orifice of the gage. Such an opening becomes so large in steep mountains that it is impractical. Some hydrologists have proposed putting gages on poles above tree crowns, but this drastically increases the effect of wind turbulence on gage catch. The forest opening serves in part to shield the gage, and usually gives a reasonable estimate of gross precipitation.

Early ideas that forests increase precipitation were most likely due to rain and snowgaging difficulties. A gage in an open field catches less water than falls because of turbulence around the orifice, whereas an identical gage in a too-small opening in the nearby forest might tend to enrich the catch due to wind funneling or splash from over-hanging trees. The result is that gages in the forest will usually catch more than gages in the open, even if gross precipitation is equal.

E. Rain intensity, duration and velocity.

1. Intensity of rain is simply the rate of delivery of water to the gage, expressed as an instantaneous rate or for any time period. Below are some examples of the highest intensities measured in the world.

Time Period (duration)	Total Depth (cm)	Location
1 hour	38	United States
2 hours	46	United States
1 day	117	Philippines
2 days	168	Taiwan
1 week	333	India
1 month	930	India
1 year	2647	Burma

The maximum possible instantaneous rate is thought to be less than 1 cm/min, although a few dubious records indicate a higher rate. The period of time that rainfall occurs is termed duration, which may represent the total storm or shorter periods of relatively uniform intensity.

2. Depth-area-duration analyses are performed for reservoir design and other purposes that require more information than point rainfall amounts and intensities. The Corps of Engineers and the Weather Service have computed the expected maximum areal delivery of rain for various areas and storm durations in the eastern U.S.:

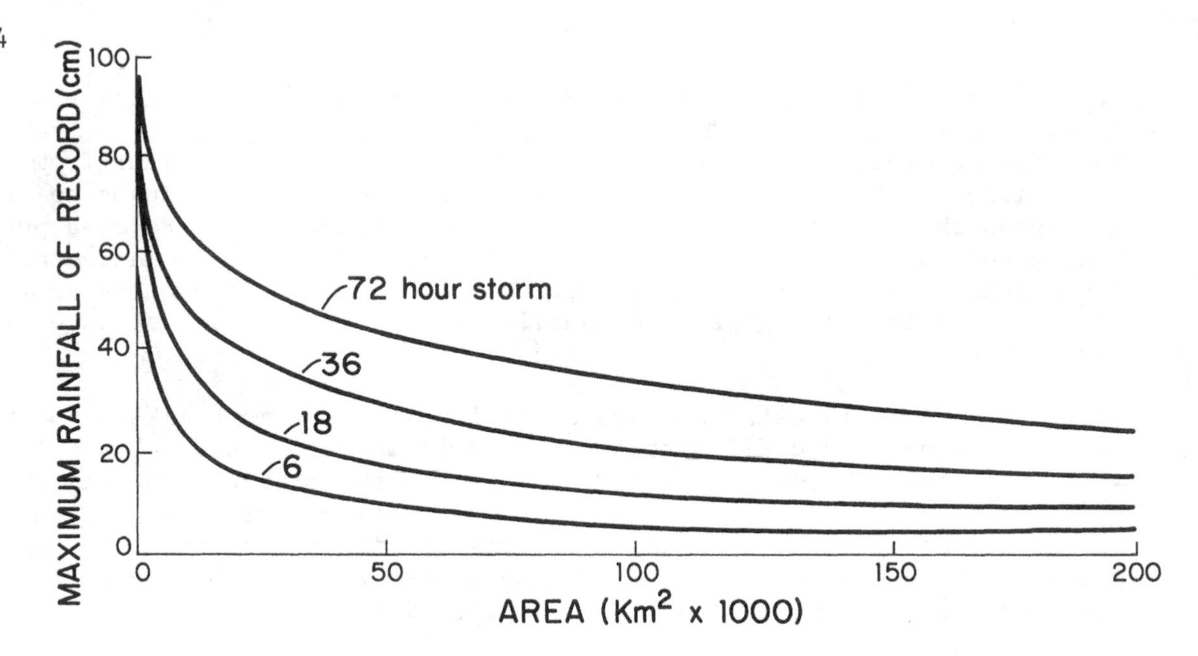

Obviously, the bigger the area, the less the average depth of rain to be expected over the whole area. Short duration storms drop off more rapidly with area than do long duration storms.

3. Rain drop velocities are particularly important in the process of erosion and sedimentation. Terminal velocity is acquired by a falling body when the force of its acceleration due to gravity is just offset by air resistance. Terminal velocities depend on drop size and shape, not so much on distance of fall. When velocity becomes great enough, drops will break into smaller ones and slow down. Drop size, velocity and rainfall intensity are related in this manner:

Type	Intensity (cm/hr)	Drop Diameter (cm)	Terminal Velocity (m/sec)
Drizzle	≤.03	≤.05	4.2
Moderate rain	.12-.38	.12-.15	5.0
Heavy rain	1.5-10.0	.25-.65	7.6

If heavy rain is delivered to bare soil, fine soil particles become entrained in water and tend to plug the infiltrating surface, causing overland flow and initiating erosion. Further details appear in Chaps 5, 7, 8 and 9.

F. Frequency analysis. In the design of dams, bridges, culverts and other structures, it is necessary to know the frequency with which storms (or other weather-related events) of a given size and intensity will be exceeded. Long-term records of weather events are scarce; few are longer than 100 years in any case. Is there a cycle in large events, or do they recur at random in accordance with their probability? Much research has failed to convince hydrologists that rainstorms occur in cycles through the years, although seasonal cycles are evident. Frequency analysis starts with a number of years of record.

1. Annual, partial and full series. If only the largest annual rainstorm, say of 24-hr duration, is selected from n years of record, and these are arrayed from the largest to the smallest, we have the annual series. Most frequency analyses are done on this series because we usually design for the larger events. If only those storms larger than 1 cm are arrayed, regardless of the number in each year, we have a partial series with a base of 1 cm. If all 24-hour rainfalls are arrayed from largest to smallest, we have the full series.

2. Return period is the average period of time expected to elapse between successive occurrences of events of given size or larger. Often misunderstood as a sort of statistical guarantee that damaging storms will appear on schedule, return period (T_r) is simply a reciprocal statement of probability (p):

$$T_r = \frac{1}{p} \qquad 4\text{-}2$$

Let's say that it has been determined that a 24-hour rain of 25 cm (or greater) will occur in about 100 out of 1000 years. Its probability of occurring in any year must be 100/1000, that is, p = 0.1. Therefore its return period is defined as 1/.1 = 10 years. But any specific 10-yr period may have several or no events of that size. It is obvious that a 10-yr record is a poor basis for estimating the size of the 10-yr return period storm, nor is the 100-year record any better for estimating the 100-year storm.

T_r is usually estimated from the annual series of storms by use of this approximation:

$$T_r = \frac{n+1}{m} \qquad 4\text{-}3$$

where n is the number of years in the record and m is the rank of any measured event. The largest storm in a 10-yr record has rank 1, the 2nd largest, rank 2, etc. Say the rank 2 storm was 12 cm in 24 hrs. Then T_r is (10 + 1)/2, or 5.5 years, for storms 12 cm or larger.

3. Use of T_r. Elementary knowledge of probability and return period is vital in management decision. Reasoning from the simplest case, let us examine how many ways an event can occur or not occur in 2 successive years:

OCCURRENCE			NON-OCCURRENCE
Once		Twice	None
OX	XO	XX	OO

Of the 4 separate ways, 3 represent occurrence (XO, OX, XX), and one represents non-occurrence (OO) in the two years. The probability that one of the 4 combinations must occur is a certainty, therefore,

$$p_2 + q_2 = 1$$

where p_2 is the probability of occurrence in 2 years, and q_2 is the probability of non-occurrence in 2 years. Now,

$$p_2 = 1 - q_2 \qquad 4\text{-}4$$

A basic rule is that the probability of non-occurrence in n tries is the product of the probabilities of non-occurrence in each try (for example, the q of not getting a head in two tosses of a coin is .5 x .5 = .25). So,

$$p_2 = 1 - (q)(q) = 1 - q^2$$

Dodging rigorous proof, we may generalize this expression by substituting n for the number of years:

$$p_n = 1 - q^n \qquad 4\text{-}5$$

Read p_n as the probability of occurrence in n years, and q^n as probability of non-occurrence in one year raised to the nth power.

> Example: An equipment shed is built to hold the 100-year snow storm. What is the chance of failure in 25 years? The probability of occurrence in any one year is $p = .01$; and of non-occurrence is $q = 1 - p = .99$. Therefore:
>
> $$p_{25} = 1 - (.99)^{25}$$
> $$= 1 - .78$$
> $$= .22$$
>
> There is a 22% chance that the shed will fail in 25 years. Suppose we are unwilling to accept that risk, but will accept a 5% chance of failure, that is p_{25} must equal .05. What return period snow storm must we design for?
>
> $$.05 = 1 - q^{25}$$
> $$q^{25} = .95$$
> $$q = .998$$

Use Eq 4-2 to compute T_r, remembering that p = 1 - q:

$$T_r = \frac{1}{1 - q} = \frac{1}{1 - .998}$$

$$= 500 \text{ years}$$

The example illustrates several things: There is always a chance of failure, and efforts to minimize it can be very costly. Furthermore, historical records of hydrologic events must be worked to the limit to estimate 100-yr return periods.

G. Calculating areal precipitation. Daily, monthly, annual or storm precipitation measured at points may be transferred to areal estimates in three ways: 1) the arithmetic mean, 2) Thiessen polygon, and 3) isohyetal methods.

The arithmetic mean is obvious. The polygon method permits the arbitrary weighting of each gage catch by the area nearest to it. It is done by connecting all gages, as plotted on a map, by a straight line. Perpendicular bisectors of each connecting line are drawn, forming polygons around each fractional area of the basin lying within that polygon. Fig 4-2 shows how basin precipitation is calculated for a single storm.

In the isohyetal method, the observer draws lines of equal precipitation, or isohyets, on the map from his knowledge of the basin topography, storm patterns, and from the amounts of precipitation measured at each gage. The basin precipitation is the sum of the weighted average precipitation between adjacent isohyets.

The polygon method is most often used but does not account well for topographic or storm pattern effects on precipitation. The isohyetal method is more accurate but its computation is time consuming because new isohyets must be drawn for each storm; it is used mostly in research and special storm analyses.

1. Missing record. What can be done if no reading was made at one of the polygon gages for the time period under study? This is a common problem usually solved by the normal ratio method. Say one month's record is missing at station X and we wish to replace it to compute weighted monthly precipitation over the basin.

$$P_x = \frac{1}{n} \left[\frac{\bar{P}_x}{\bar{P}_1} P_1 + \ldots. + \frac{\bar{P}_x}{\bar{P}_n} P_n\right] \qquad 4\text{-}6$$

The estimate of the missing value P_x is computed by using the ratio of that station's average monthly precipitation ($\bar{P}_x$) to each of the other station's monthly averages ($\bar{P}_n$), and carrying out the arithmetic in Eq 4-6. Of course, there must be simultaneous records of precipitation during other months to use the method.

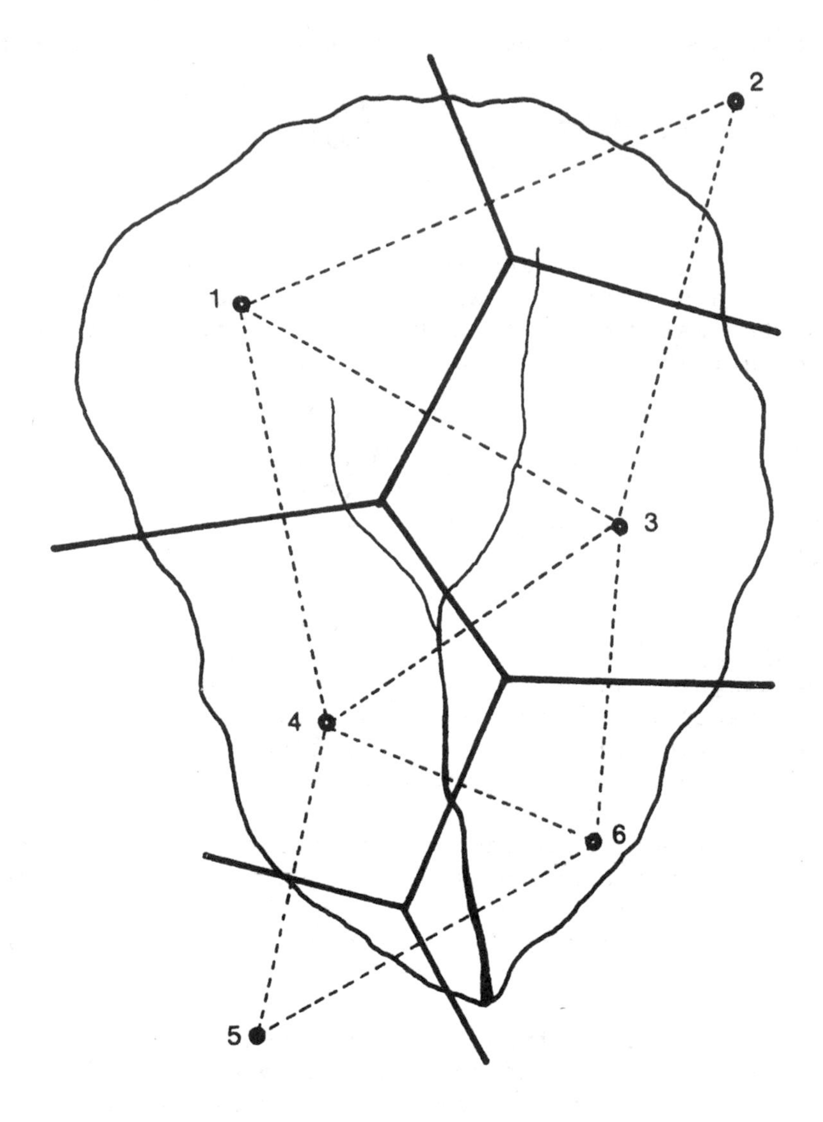

Rain Gage Station	Rainfall (cm)	Fraction of Basin in Polygon	Weighted Rainfall (cm)
1	16	0.30	4.8
2	15	0.08	1.2
3	14	0.28	3.9
4	12	0.21	2.5
5	10	0.01	0.1
6	5	0.12	0.6
TOTALS	72cm/6 = 12cm	1.00	13.1

Fig 4-2. The diagram shows how polygons are constructed for estimating weighted rainfall. While the arithmetic mean indicates 12 cm over the basin, the polygon method gives a more accurate estimate of 13.1 cm. If the volume of rain is important, we must know the basin area.

H. Precipitation water quality. Even pure rainwater has a small amount of CO_2 dissolved in it, giving a weak acid reaction. Augmented by humic acids leached from soil, the carbonic acid in rainfall has been sufficient in geologic time to wear away many rock types, particularly sedimentary limestones. A recent concern about "acid rain" relates to the stronger acids, SO_2 and NO_x. Among the sources of SO_2 and NO_x in the atmosphere are burning of fossil fuels, industrial chemical plants and volcanoes. The erosion of stone buildings and monuments, the leaching of soil nutrients, the acidification of lakes, and the incidence of lung disease in humans are attributed partly to such air pollution. The most obvious form of damage occurs at metal smelting centers around the world; a famous example is Copper Basin, Tennessee, where 100 years of copper smelting by a sulfuric acid method totally denuded 100 km^2 around the plant, making re-establishment of vegetation and the restoration of soil fertility difficult.

In northern latitudes shallow lakes and soils which are naturally high in humic acids and low in bases are reportedly susceptible to sterilization by increased precipitation acidity. Lakes and soils in temperate climates are not so easily altered. Early results of "acid rain" studies indicate that natural sources of acidity in the atmosphere constitute a basic environmental problem that is aggravated locally by man's activities, as in the case of Copper Basin.

Some agents in precipitation (nitrates, calcium, trace elements) are beneficial to soil fertility and plant growth.

I. Snowpack and melt. Snow hydrology is too broad a subject to do justice to in this outline. The student is referred to various bulletins of the Forest Service, U.S. Department of Agriculture, particularly the Rocky Mountain and Pacific Southwest Forest Experiment Stations.

About 75% of western U.S. water supplies come from forests and wildlands, much in the form of snow. Snowfall and snowpack conditions are too spatially variable to measure on a routine basis. But comparative measurements, or indices, are acquired at fixed snow courses which are visited at intervals during winter to determine the amount of water that may be expected during spring melt periods. Snow course measurements in the mountains are plotted against subsequent riverflow in the valleys. After many years of record, these relationships have become reliable enough to allocate water under appropriative rights to users several months in advance. This allows water users to plan and operate more effectively.

1. Snowpack measurement. Several things may be measured along the snow course, a permanently staked line about 100 m long chosen to be representative of the site. The easiest is snowpack depth, which can be read from the air if the stakes are calibrated. Normally the course is visited at monthly intervals, at which time depth, snow density and water equivalent are determined by drilling out a core of the snowpack with a snow sampling tube. Density of new snow ranges from 0.05 to 0.20 g/cm^3, the former for cold dry snow, the latter for settled wet snow. Water equivalent is the

depth of water produced by the melted snow, measured in cm. Depth, density and water are related, remembering that 1 cm^3 of water weighs 1 g:

$$\text{Water equivalent} = (\text{snow density})(\text{snow depth})(1\ cm^3/g)$$

As the snowpack ages (ripens), snowflakes undergo deformation and compaction into a series of forms well known to skiers, finally taking a density near 0.9 in a peculiarly blue form of ice (glacier). "Well-ripened" snow has a density of 0.60 or more and if late in the season may have developed low "quality." Snow quality is the percentage of the snowpack that is ice: Fresh wet snow has a quality of 95% or more (5% of the snow is water) whereas a draining snowpack's quality may be only 60% (40% water). Under conditions of low quality, relatively small additions of energy by warm air or rain will cause rapid melting and possibly spring floods. To anticipate such events, snow quality is measured by calorimetry and reported to flood forecasting centers.

Example: A sample of snow from a snowpack 60 cm deep has a volume of 15000 cm^3 and weighs 2 kg. When mixed with 7 kg of 32°C water, the mixture takes a final temperature of 8°C. Determine snow density, water equivalent and snow quality.

Snow density = $(2000g/15000\ cm^3) = 0.133 g/cm^3$

Water equivalent = $(.133g/cm^3)(60cm)(1cm^3/g) = 8.0cm$

The specific heat of water is 1 cal/g/°C, latent heat of fusion is 80 cal/g, and the snow is initially at 0°C.

Heat required to raise melted snow to 8°C:

$$(2000g)(1\ cal/g/^\circ C)(8^\circ C) = 16000\ cal$$

Heat furnished by added water:

$$(7000g)(1\ cal/g/^\circ C)(32^\circ - 8^\circ C) = 168000\ cal$$

The heat balance:

$$168000\ cal - 16000\ cal = (80\ cal/g)(\text{ice content})$$

$$\text{Ice content} = 1900g$$

Snow quality = 1900g of ice/2000g of snow

= 0.95

2. Avalanches are frequent and sudden in high mountains, trapping skiers, destroying buildings, blocking highways and cutting swaths through forests. Efforts to control or predict avalanche hazards are highly developed in Switzerland and in some parts of

the western U.S. The student is referred to U.S. Forest Service Research Paper RM-19, "A manual for planning structural control of avalanches" (1966), for information on avalanche problems.

3. Snowmelt. One of the objectives of snowpack management might be to delay the time of melting so as to sustain streamflow into the summer. But only within narrow limits can man influence the sources of heat energy for melting snow:

> Turbulent exchange of sensible heat with the air cannot be controlled in management, but accounts for most snow melt.
>
> Short and long wave radiation from sun and sky can be controlled to some extent, as in the case of spreading coal dust on snow to trap radiation and speed melt rate, or in the case of forest management to shade snow and delay melt.
>
> Heat of condensation of water vapor onto the snowpack cannot be controlled. One gram of condensed water yields sufficient heat to melt 7.5 g of ice, that is, the latent heat of condensation divided by the latent heat of fusion. The amount of condensation into snowpacks is not well known but it is doubtful that condensation heat is a large cause of snowmelt.
>
> Rain heat accounts for some melting but it is not under managerial control.
>
> Heat conduction from the soil cannot be controlled but is thought negligible as a source of heat to melt snowpacks.

The last three do not account for appreciable melting of snowpacks, but all these factors vary from region to region and season to season. Snowpacks resist melting because their albedo (the reflectivity of short wave radiation) is very high (almost 95% in new snow) and long wave emission serves to lower the temperature. Furthermore, the heat conductivity of snow and ice is very low and convection within the pack is probably negligible. Not until the warm winds of spring arrive, and the albedo of the aging snow decreases, does environmental heat gain access to the snowpack.

During winter the snowpack is reduced somewhat by sublimation. Estimates of snowpack evaporation vary widely but it is thought that snow shaded by forest cover loses appreciably less evaporation than snow in direct sun. Fig 4-3 shows how silviculture might be planned to reduce evaporative loss and delay snow melt; such prescriptions are not yet applied as a practical matter but are recommended.

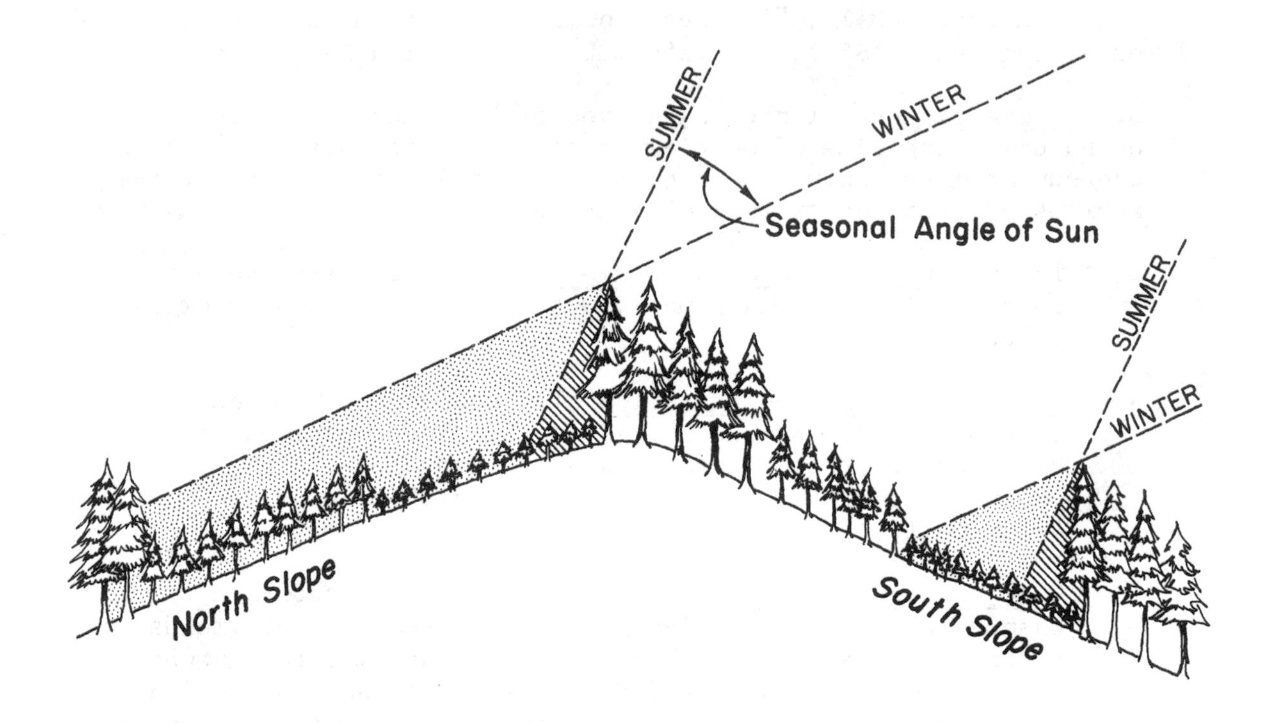

Fig 4-3. Progressive cuts may be performed in snow country to establish a "wall-and-step" forest for the purpose of increasing snow accumulation and delaying snowmelt into late spring (from Anderson 1963).

Another technique to delay snowmelt consists in judicious placement of snow fences to increase snowpack depth on favorable sites. Deep snowpacks resist melting and provide a substantial percentage of late summer streamflow from high mountain areas of the West. Pilot tests are underway to determine the feasibility of improving mountain streamflow regimen in some areas.

Further readings.

Anderson, H. W. Managing California's snow zone lands for water. U.S. Forest Service, Research Paper PSW-6, 28 pp., 1963.

Frutiger, Hans and M. Martinelli, Jr. A manual for planning structural control of avalanches. U.S. Forest Service Research Paper RM-19, 68 pp., 1966.

Gartska, W. U. Snow and snow survey. Section 10 in V. T. Chow (Ed), Handbook of Applied Hydrology, McGraw-Hill, N.Y., 57 pp., 1964.

Gilman, C. S. Rainfall. Section 9 in V. T. Chow (Ed), *Handbook of Applied Hydrology*, McGraw-Hill, N.Y., 68 pp., 1964.

Lee, R. *Forest Microclimatology*. Columbia Univ. Press, N.Y., 276 pp., 1978.

Penman, H. L. *Humidity*. Chapman and Hall, Ltd. for Reinhold Publishing Corp., N.Y., 71 pp., 1955.

Schroeder, M. J. and C. C. Buck. *Fire Weather*. U.S. Forest Service, U.S.D.A., Agriculture Handbook No. 360, 229 pp., 1970.

U.S. Dept. of Agric. *Field Manual for Research in Agricultural Hydrology*. U.S. Gov't. Print. Ofc., Wash., D.C., Agriculture Handbook No. 224, Chaps. 1 and 3, 98 pp., 1976.

Problems.

1. How many calories are required to convert 1 g of water at 20°C into steam at 100°C?

2. What is the probability of a 100-yr storm being equaled or exceeded in the next 100 years? If only a 1% risk of failure is acceptable for a structure that is to be used 50 years, for what return period must the structure be designed?

3. Compute the missing values in the following annual rainfall record (in cm) at 3 associated precipitation stations:

Gage	------ Annual Precipitation (cm) ------							
	1951	1952	1953	1954	1955	1956	1957	1958
1	112	121	108	122	141	125		131
2	102		98	108	133	117	115	122
3	116	127	111		144	131	126	133

4. On a map of a basin to be handed out in class, determine the basin precipitation first by the arithmetic mean and then by the polygon method.

5. A standard rain gage is to be located in a forest with trees 25 m tall. The orifice is 1.37 m above the ground. The ground slopes evenly at 20° from horizontal. Design the clearing that must be made if the usual 45° clear line of sight is required in all directions from the orifice. (Solve graphically.)

6. A 2-m snowpack with a snow density of 0.15 g/cm^3 receives 5 cm of 10°C rainfall. At the beginning, the snowpack is at 0°C with a snow quality of 0.90. Assume that no drainage takes place before the end of the rain and calculate the new snow quality to the nearest 0.01 unit.

CHAPTER 5

SUBSURFACE WATER

A. Introduction. About 75 percent of all precipitation in temperate climates enters the surface of the soil to become soil water, either soil moisture in unsaturated soil, or ground water in saturated soil and rock. As with vapor in the air, soil water is invisible and its behavior seems a mystery to most people.

We will begin by classifying the regolith into the zone of aeration (soil moisture) and the zone of saturation (ground water). The water table separates these zones (Fig 5-1). The root systems of most plants are restricted to the zone of aeration because air (particularly oxygen) is essential to root growth and function. Because most crops depend on the top meter of the zone of aeration, soil scientists classify soils differently by concentrating on the solum, which is the part of the regolith that serves as a medium for plant growth.

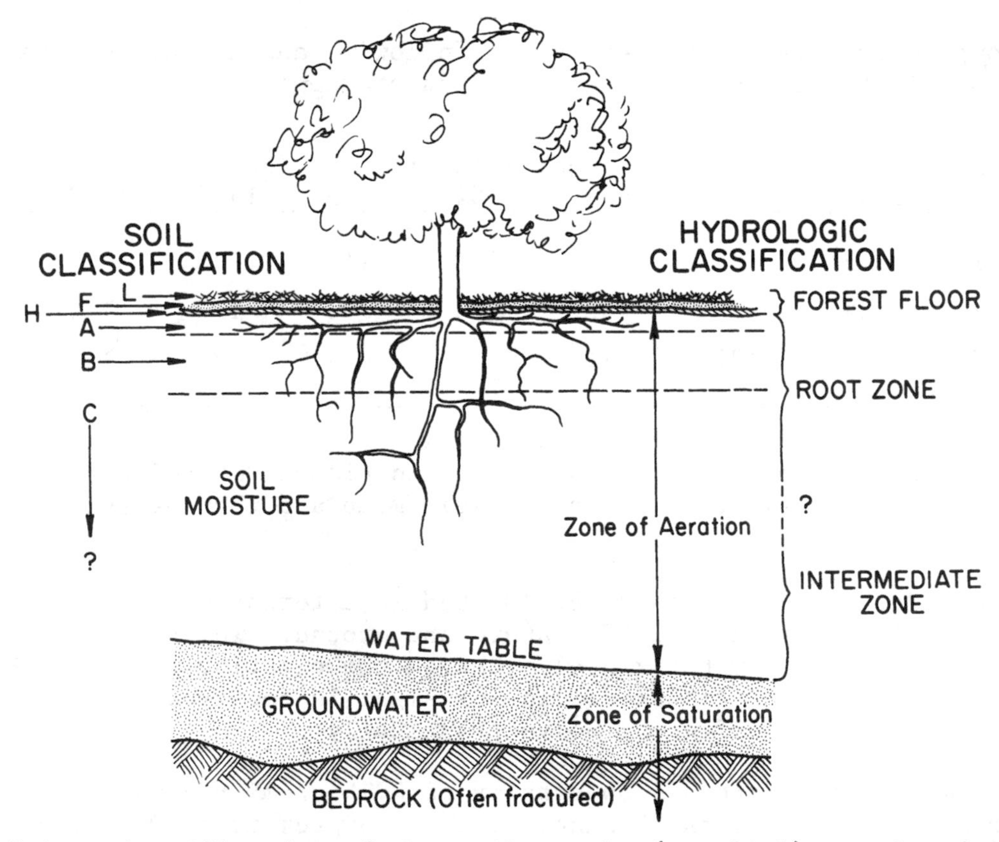

Fig 5-1. Classification of the soil mantle (regolith) as viewed in soil science and in hydrology. L, F and H are the litter, fermentation and humus layers, respectively.

B. The soil mineral complex may be divided into five particle size classes: gravel (> 2mm), coarse sand (0.2 - 2mm), fine sand (0.02 - 0.2mm), silt (0.002 - 0.02mm) and clay (< 0.002mm). Gravels and sands may be separated by sieving but clays and silts must be separated by differential settling rates in still water. Silts settle in a few minutes but fine clays require days, even weeks. The texture of soil refers to the relative amounts of the particle classes it contains. The tendency of soil particles to aggregate into crumbs, columns, clods or blocks gives soils their structure. Soil structure is affected by texture, organic matter, mineral type and biological activity (particularly fungi and earthworms). A crystalline, fine sand has no structure. If clay or colloidal organic matter is added, aggregation may occur, changing the distribution of pore sizes and often increasing total porosity. Small pores remain within the aggregates but larger pores form between aggregates. The total pore space of a given soil may vary from as little as 35% when poorly aggregated to as much as 65% when well aggregated. In summary, texture and structure affect the distribution of pore space, which in turn affects infiltration, detention storage, water movement, soil moisture storage and plant growth, all important hydrological and biological processes. Soil physical factors are thoroughly outlined in Brady (1974).

C. Soil moisture. Pores within the zone of aeration contain liquid water and soil air (the latter may differ from atmospheric air). A volume of soil (V) will be part solid matter (V_s), part water (V_w) and part soil air (V_a):

$$V = V_s + V_w + V_a \qquad 5\text{-}1$$

1. Bulk density is an expression of the dry mass of soil occupying a given space in the soil mantle. Bulk density (B) is mass per unit volume of dry soil, the volume representing the exact space the soil occupies in the field (V in Eq 5-1).

$$B = \frac{\text{Mass of the dry soil in g}}{\text{In situ volume in cm}^3} \qquad 5\text{-}2$$

In the metric system, mass and weight may be taken as numerically equal at the earth's surface, and the number of grams represents weight, mass and cm^3 of water all at once. A sample of known volume is cut from the soil, oven dried at 105°C, weighed in grams, then divided by the volume in cm^3. In the English system, bulk density is dimensionless, representing the density of soil compared to density of water:

$$B = \frac{(\text{Weight of soil in lbs})/(\text{In situ volume in ft}^3)}{62.4 \text{ lbs/ft}^3} \qquad 5\text{-}3$$

If the weight of water per ft^3 were not used in Eq 5-3, then B in Eqs 5-2 and 5-3 would not be numerically equal. It is much simpler to do soil moisture computations in the metric system.

B ranges from 0.5 in loose surface horizons to 1.8 in compacted sands; forest subsoils generally range from 0.9 to 1.3.

2. Particle density (numerically equal to specific gravity of soil particles) is larger than bulk density unless porosity is zero. Most soil particles are quartz, feldspars or silicates which have a particle density of about 2.6 g/cm^3. A piece of solid granite has the same density as the particle density of soil derived from it.

3. Soil porosity (ξ) is the fraction of total soil volume that is pore space:

$$\xi = \frac{V_w + V_a}{V} \tag{5-4}$$

4. Soil moisture is usually based on the amount of water lost from a sample when dried in an oven at 105°C for 24 to 48 hrs. Using this procedure, the volume of water (V_w) includes the tiny amount of water vapor present in the soil. It follows from Eq 5-1 that the moisture content as a fraction of bulk soil volume (Θ_v) is:

$$\Theta_v = V_w/V \tag{5-5}$$

Soil scientists often prefer to express moisture content as a fraction of dry soil weight (Θ_w):

$$\Theta_w = W_w/W_s \tag{5-6}$$

where W_w is the weight of water in a soil of dry weight W_s.

The relationship between bulk density (B), moisture content by volume (Θ_v) and by weight (Θ_w) is:

$$\Theta_v = B\ \Theta_w \tag{5-7}$$

5. Soil saturation occurs when all pore spaces are filled with water. The saturation water content should be equal to the porosity. However, in field soils there are always pockets of gas trapped within the pores, as much as 5 to 8 percent of the total volume. Because of this, water tables sometimes rise or fall slightly with barometric pressure changes.

Volumetric soil water content is used to calculate an equivalent depth of water over the land. The use of water content by weight can be misleading because soils with the same water content by weight may have quite different water volumes. The expression of soil water content as an equivalent depth of water is essential when considering changes in basin storage (ΔS), or to relate precipitation, soil water, evapotranspiration and streamflow in the same units of measurement.

What is dry soil? An air dry soil is one which has come to moisture equilibrium with the surrounding atmosphere. It will gain or lose slight amounts of water with a change in air humidity. On the other hand, an oven dry soil by definition has been dried at 105°C until it reaches constant weight. Even an oven dry soil is not completely devoid of water because water molecules are strongly bonded in the colloidal material (clay lattices, organic matter) and to mineral surfaces. At temperatures up to 500°C these bonds are broken and the water is vaporized. This water is not available to plants nor is it a part of normal hydrologic processes. We therefore determine all soil water contents on the basis of soil dried at 105°C.

Example: A soil sample had an in situ volume of 20.0 cm^3 and weighed 30.6 g. After oven drying, it weighed 25.5 g. Determine B, Θ_w, and Θ_v.

$$B = 25.5g/20.0cm^3 = 1.275g/cm^3$$

$$\Theta_w = (30.6 - 25.5)g/25.5g = 0.200$$

$$\Theta_v = (30.6 - 25.5)g/20.0cm^3 = 0.255g/cm^3$$

$$\text{Or: } \Theta_v = B\,\Theta_w = (1.275g/cm^3)(0.200) = 0.255g/cm^3$$

Since 1 cm^3 of H_2O = 1 g, moisture content by volume may be multiplied by 1 cm^3/g to give a dimensionless fraction for use in any system of units. In this example, how many inches of water are in an 18-inch layer of soil?

$$\text{Inches of } H_2O = (\Theta_v)(\text{Profile depth})$$

$$= (0.255\ g/cm^3)(1\ cm^3/g)(18\ in)$$

$$= 4.59 \text{ inches of } H_2O$$

D. Soil moisture measurement. Because of heterogenous soils, differential root concentrations and slope positions, many measurements at different points are necessary before the average moisture content of a watershed can be estimated with any useful accuracy. Even the best method cannot attain accuracy to within 0.01 Θ_v in the field. The following methods are most commonly used.

1. The gravimetric method makes use of Eq 5-7. Both B and Θ_w must be sampled, but the product varies much more than B and Θ_w individually. Soil samples may either be taken "undisturbed" (occupying original volume), in which case both B and Θ_w may be measured on the same large sample, or in two separate operations, a large B sample and a small Θ_w sample, the latter secured by a small steel tube driven deeply into the earth. Both have disadvantages: Undisturbed sampling is site destructive, time consuming and can only be done in shallow layers. Separate sampling reduces site destruction and cost by delaying the sampling of B

until the end of the study, but at the same time greatly increases variation of the estimate. Measuring basin water storage by either method is impractical because the standard error of estimate is too large to be hydrologically useful. Gravimetry is useful in the greenhouse, in surface soil layers and as a check on other methods.

2. The electrical conductance method is based on the fact that the electrical conductivity of a material varies with its moisture content. The method employs porous blocks, constructed of nylon, fiberglass or gypsum, that have two imbedded electrodes. When placed in the soil, the blocks take up or release water depending on the potential gradient between the block and surrounding soil. Resistance to electrical current within the block is related to its moisture content. Because the block's moisture is in equilibrium with soil moisture potential, not with moisture content, electrical conductance methods are better used to indicate plant growth conditions than to measure moisture content of soil. Although blocks permit repeated measurements at one point, they deteriorate rapidly in the soil.

3. The neutron scattering method uses radioisotopes to measure soil moisture content. The device employs a radioactive source of fast neutrons. The neutrons have a mass almost identical with hydrogen. Neutron speed is moderated and the path deflected more by hydrogen than any other element in most soils. The slow neutrons are absorbed in a detector, counted electronically and related by calibration to the density of the hydrogen field near the source. Because hydrogen in the soil is directly related to water, measuring the hydrogen density is equivalent to measuring water density (g/cm^3). The neutron scattering method permits repetitive, nondestructive and precise measurement of moisture and moisture change around a probe lowered into a tube in the ground. The volume of soil sampled is large, approximately $0.04\ m^3$, and depths up to 12 m may be reached without severe disturbance of the soil horizons, unless the soil is stony. The speed and convenience of the neutron scattering method allows more measurements, more frequently, to estimate a basin's change in storage.

E. Energy of water in soil. Energy involves forces of attraction among bodies as well as thermal and chemical energies. The attraction of water molecules for each other is called cohesion and the attraction between soil particles and water is called adhesion. Cohesive and adhesive forces operate to produce negative pressures (tension or suction forces) in the water between soil particles. As water drains from pores, a curved air-water interface (called a meniscus) develops due to the surface cohesion (tension) of water. At the same time the films of water around the particle become thinner, bringing the water under stronger and stronger adhesive force.

Water in small pores is held more strongly than water in larger pores. The pressure in the soil air is atmospheric (assuming the air pockets are connected), but pressure inside the water films is less than atmospheric (negative).

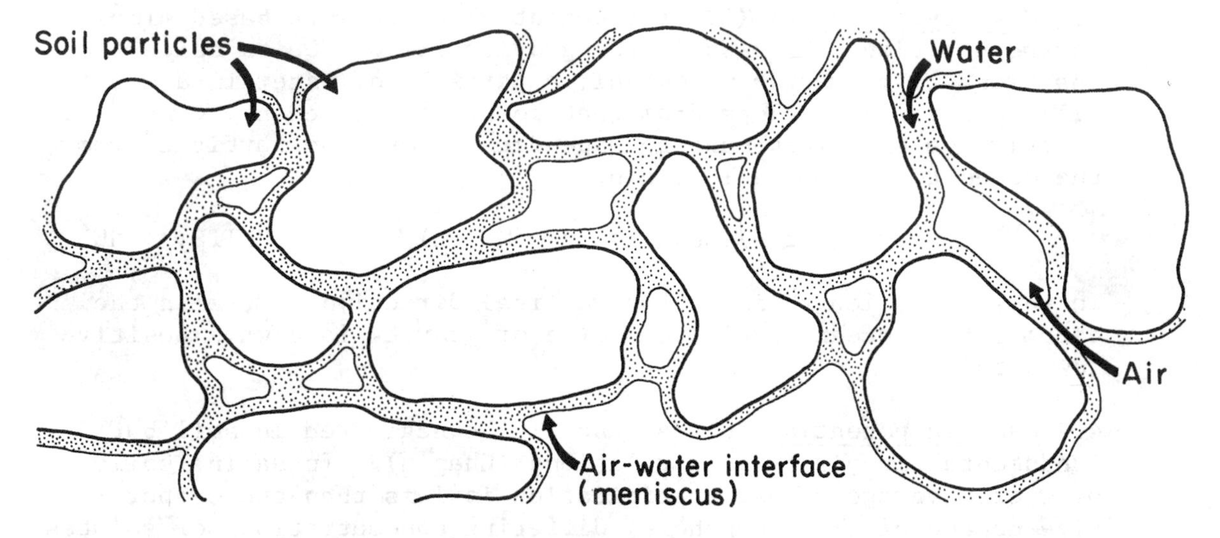

If there were no negative forces holding moisture in unsaturated soil, the water would drain freely out under the force of gravity. As it is, some water drains out after being detained for a while, and some is retained for a long time. The energy status of soil water is confusing only if we try to ignore the elementary physics involved. Just how is energy related to water in soil?

1. Mass is simply an amount of matter, water in this case. To measure mass, we weigh it on a scale relative to standard weights. But weight does not equal mass:

Mass = Weight/Acceleration due to gravity = W/g   UNITS: M

The units of weight (W) are different:

$W = Mg = (M)(LT^{-2})$ UNITS: MLT^{-2}

The mass of an object is the same everywhere; its weight depends on the acceleration due to gravity (LT^{-2}), which depends on where the object is in the gravity field. Weight is a force (F), or mass times acceleration:

$F = Mg$ UNITS: MLT^{-2}

2. Pressure potential (P) is a potential to do work, that is, to expend a force over a distance:

$P = (\text{Force})(\text{Length}) = (Mg)(L)$ UNITS: ML^2T^{-2}

Under the water table P is positive and in unsaturated soil P is negative. The + or - is a convention; that is, we conveniently set P = 0 at the standard pressure of the atmosphere (1.013 bar), then measure pressure potential relative to that datum. In soil, negative pressure potential (-P) is called matric potential (the matrix is the network of soil particle surfaces to which water adheres), and in the positive range, pressure (head) potential, or just head.

3. Gravity potential (Z) is a potential to do work based simply on the position of a mass in the gravity field. Quite aside from its pressure or matric potential, a particle of water in a mountain will do work as it drops from that point to the ocean. Conversely, exactly the same work must be expended to move the particle from the ocean to the mountain again.

$$Z = (\text{Force})(\text{Length}) = (Mg)(L) \qquad \text{UNITS:} \quad ML^2T^{-2}$$

In this case, length is in the vertical direction and, with the datum set at ocean level, the force of gravity is always positive downward.

4. Osmotic potential (0) may usually be neglected in soil but is fundamental in plant-water relations (Chap 6). In saline soils, or where average annual precipitation is less than the evaporative demand of the atmosphere, differing concentrations of solutes from place to place add a chemical potential to P and Z. Water will be forced to move from areas of low solute concentration to areas of high solute concentration (the purer water "tries to dilute" the impure water by moving into the stronger solution). Again the potential for work is:

$$0 = (\text{Force})(\text{Length}) = (Mg)(L) \qquad \text{UNITS:} \quad ML^2T^{-2}$$

The force may operate in any direction, to or from a given particle of soil water.

Thermal potentials also exist in soil water but the minor forces exerted by differences in temperature from point to point will affect either matric potential (-P) or osmotic potential (0), therefore no additional term is needed in hydrology. Thermal water movement in soil is thought to operate mainly in the vapor phase since heat raises or lowers the vapor pressure (Chap 4) in soil air.

Now we may define the total potential energy (Ψ) of a liquid water particle anywhere in the soil. It is the sum of the three component potentials above:

$$\Psi = P + 0 + Z \qquad \text{UNITS:} \quad ML^2T^{-2} \qquad 5\text{-}8$$

It is convenient to express these potential energy components in energy per unit weight of water. The result is simply length, usually measured in cm:

$$\Psi/Mg = P/Mg + 0/Mg + Z/Mg \qquad \text{UNITS:} \quad ML^2T^{-2}/MLT^{-2} = L$$

We will continue to use the same symbols as in Eq 5-8 for the water potential balance, remembering that it is measured and expressed in length (cm). Gravity potential is easily calculated for any point in the soil as a length in cm above some datum, usually set conveniently well below the soil system under analysis. Osmotic potential can be measured cryoscopically or electronically, but it is usually neglected in soil computations. Pressure potential per unit weight (P) is measured in manometers (piezometers) in the positive range, and tensiometers in the negative range (below atmospheric pressure potential, which is approximately equivalent to the potential exerted by a column of water 10 m high). A manometer is a simple tube inserted into a saturated soil volume to measure the height to which water in the vertical part of the tube will rise against the pressure of the atmosphere.

5. Tensiometers are useful to clarify potential energy of soil water (Fig 5-2). The water in the U-tube comes to equilibrium potential with the water in the soil because water moves in or out of the porous ceramic cup. The cup must be in close contact with the films of water around the soil particles. The quanity -P in cm is a direct measure of the matric potential, and the quantity Z in cm is a relative measure of gravity potential at the center of the cup. Neglecting osmotic forces, the total potential Ψ is shown as Z - P. If the water level were just opposite the cup, the soil would be saturated to the level of the cup and P would be zero. If the water table rose above the cup, P would be positive (pressure potential) and the tensiometer would be operating as a manometer. In commercial tensiometers, mercury replaces water in the U-tube; cm of mercury must be multiplied by 13 because mercury is 13 times the weight of water.

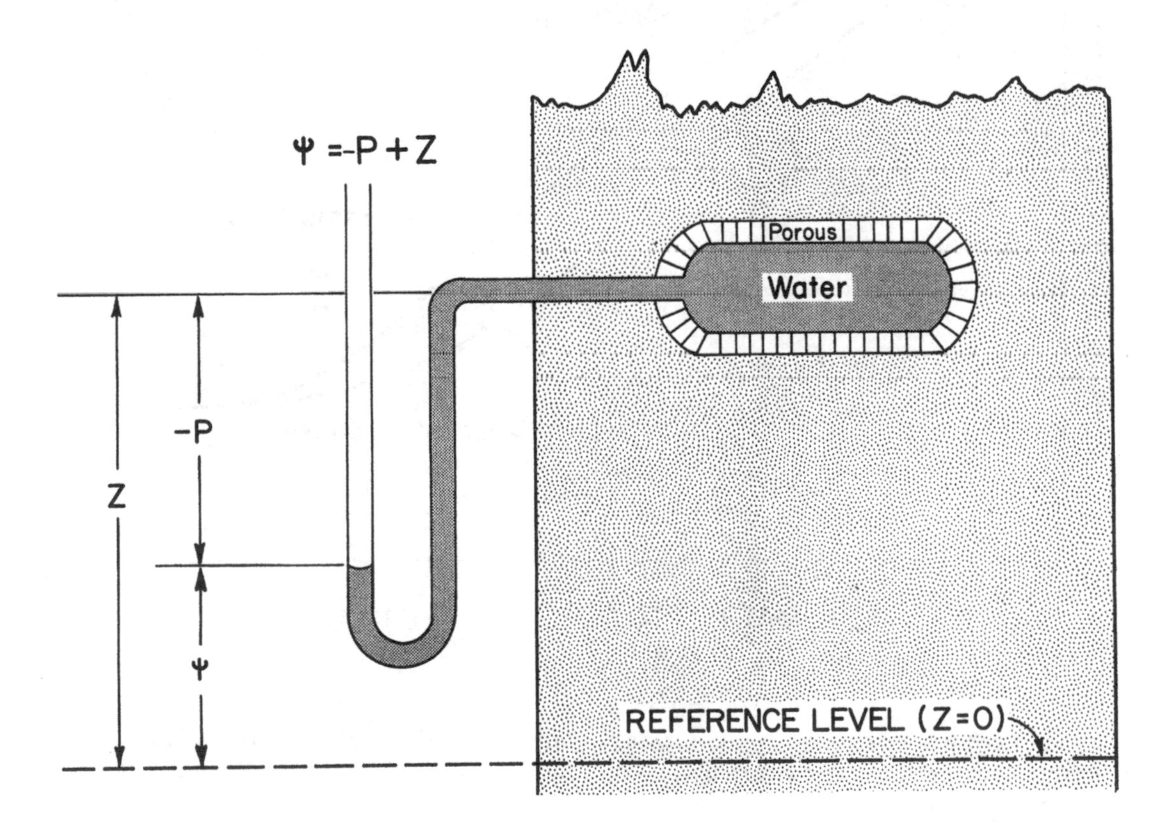

Fig 5-2. Soil water potential (neglecting osmotic potential) is measured by a porous cup permeable to both water and solutes.

6. Moisture characteristic curves. Each soil, depending on its texture and pore size distribution, exhibits a characteristic retention of moisture over a range of matric potentials. The typical shape of the relation is shown in a diagram. As matric potential decreases (more negative) smaller and smaller pores are drained and the films of water around particles become thinner. If water is fed to the soil, thus increasing matric potential, the smaller pores refill but the larger pores resist absorption because the curvature of the meniscus is too weak to pull water in. Therefore the figure shows a hysteresis loop having a desorption and an absorption phase. The phenomenon of hysteresis, meaning literally that historical moisture contents are affecting current behavior of water, is most difficult to account for in soil water behavior. It would be convenient if there existed for each soil a unique relation between matric potential and moisture content; but not so. Nevertheless, in irrigation and other hydrologic work, uniqueness is often assumed because the desorption (drainage) phase is of most interest.

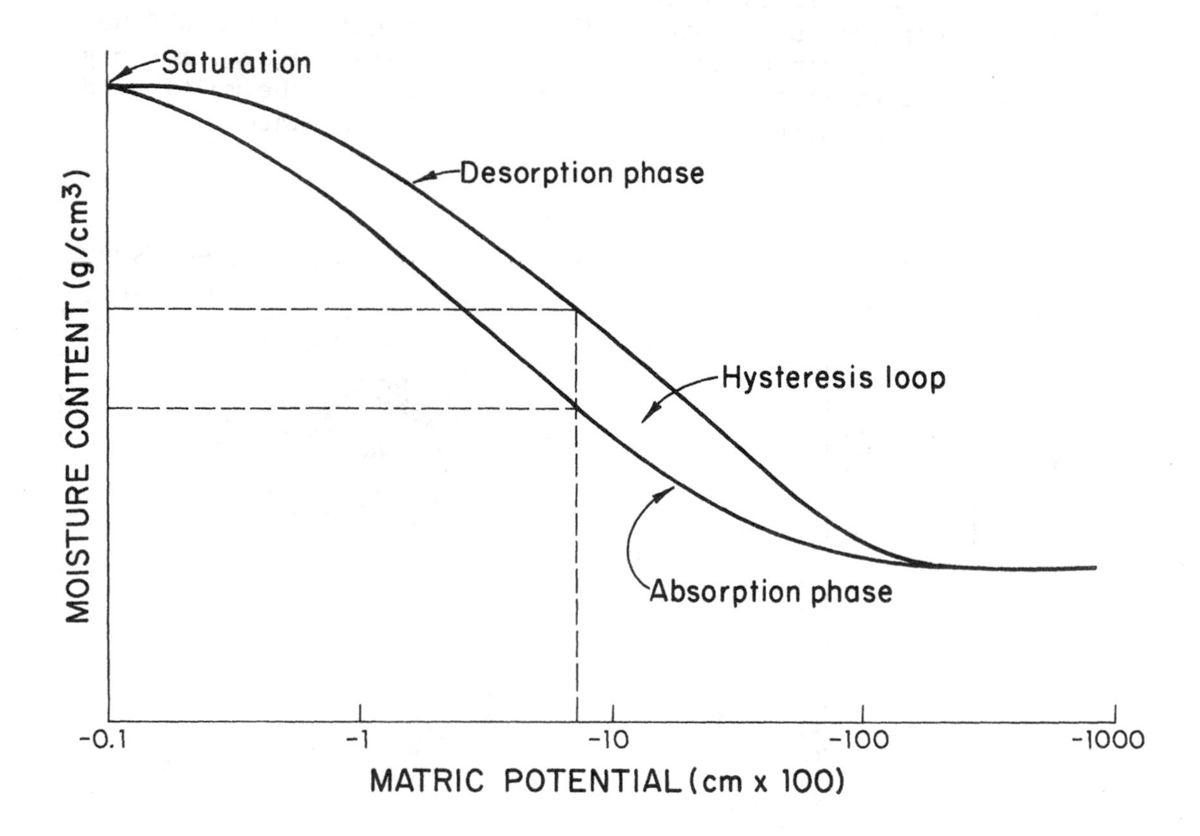

F. Soil water terminology is better developed than hydrological terminology but a number of troublesome terms require careful definition.

1. Field capacity (FC) is a useful approximation in irrigation practice and in general description but is not rigorously defined. Usually it means the moisture content of field soil 2 to 3 days after a soaking rain or heavy irrigation. Theoretically it is

"the moisture a particular soil can hold against gravity," but that is imprecise for a number of reasons. Draining moisture approaches an equilibrium moisture potential that depends on both gravity Z and matric potential -P (Sec E and Eq 5-8). The latter is characteristic of each soil, but Z depends on where the soil is located in the soil mass, that is, whether along the stream or upslope, whether in the surface layers or near the water table, and finally whether the location is receiving drainage from above or only yielding drainage downward. The term field capacity is an effort to define an arbitrary point on a time-vs-drainage curve, the moisture content where "rapid" drainage has ceased. That point is fairly clear in sand but quite unclear in clay loam:

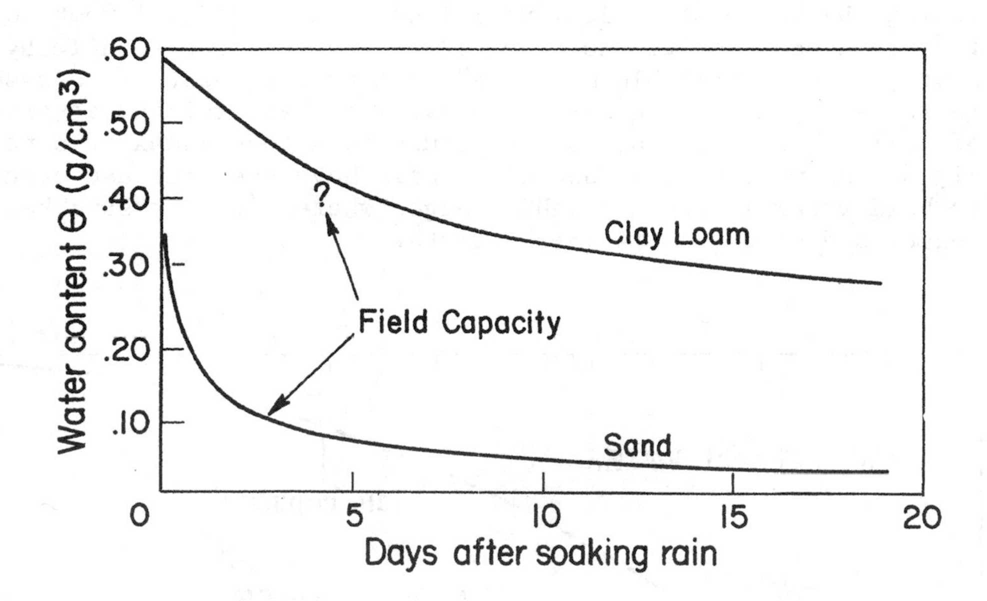

In theory, we may accurately define ultimate field capacity as the moisture content when (during desorption) the cohesive-adhesive force in the moisture films just cancels the gravitational force, that is, Z - P = 0. In large draining soil masses, such as a watershed, the equality of Z and P would occur only at infinity. FC is not used in this ultimate sense, but in the short-term sense. The value FC assumes in the field depends not only on soil texture but also on time and position in the soil mass. In addition, it depends on how deeply the soil has been wetted by a "soaking rain" and how rapidly evapotranspiration is taking place. The related terms below are useful in description but also are not well defined:

2. Gravitational water is water drained from soil before FC is reached. The term is equivalent to detention storage of water, implying that large-pores detain, but do not retain, percolating water. The implication that movement under gravity suddenly stops at FC is an approximation, useful but misleading.

3. Permanent wilting percentage is the moisture content of soil when plants growing in it can no longer maintain turgid leaves. When soil around roots remains at this moisture content for some

time, relatively permanent wilting of the leaves occurs and the plant may defoliate, die or go dormant. PWP is soil specific but is generally assumed to occur at about the same matric potential in all soils, around (-P = 150 m), or -15 bars of pressure. Some plants, for example the southern pines, can temporarily stand -50 bars and still recover. Because only tiny volumes of soil moisture are removed between -15 and -50 bars, PWP may be regarded as almost invariant for different plants growing on a particular soil type.

4. Available moisture is defined as the water held between FC and PWP, and is thus an index of the water available to plants. The amount of water stored in the root zone is a key factor in determining the value of agricultural land. The perennial roots of forest vegetation, the deep penetration of the soil mantle by tree roots, makes "available moisture" a difficult quantity to assess in forestry, except in nursery beds. As a general characteristic of soils, however, available moisture is a good index of site. Fig 5-3 shows that silt and clay loams have the greatest capacity to hold water in the available range, whereas coarse sand has the least, and pure clay is intermediate.

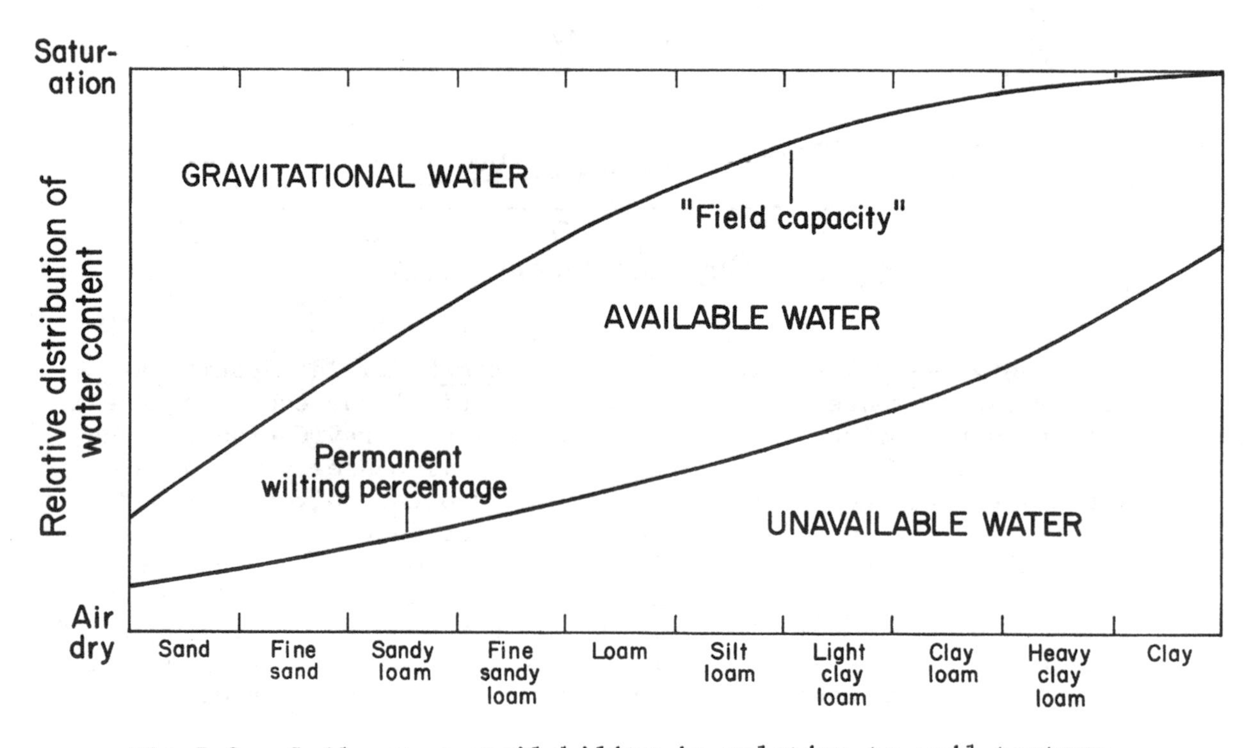

Fig 5-3. Soil water availability in relation to soil texture.

5. Hygroscopic moisture forms thin films, perhaps a few molecules thick, around soil particles below the PWP. Liquid water movement has virtually ceased at this moisture level and further vapor movement is of minor consequence in the hydrologic cycle. Hygroscopic moisture is unavailable to plants.

6. <u>Retention storage</u> is the sum of available and hygroscopic moisture. Not strictly separable from detention storage, retention storage is a quantity that presumably must be satisfied by rainfall before a soil can yield water to groundwater or streamflow.

When the above terms lead to confusion, a retreat to the water potential terminology outlined in Sec E will often resolve the difficulty. For example, soil scientists sometimes specify FC as equal to the water content of a sample of soil drained to equilibrium under -100 cm matric potential (-60 and -333 are also used). So determined, FC is a characteristic moisture index of the soil but not an accurate picture of field conditions.

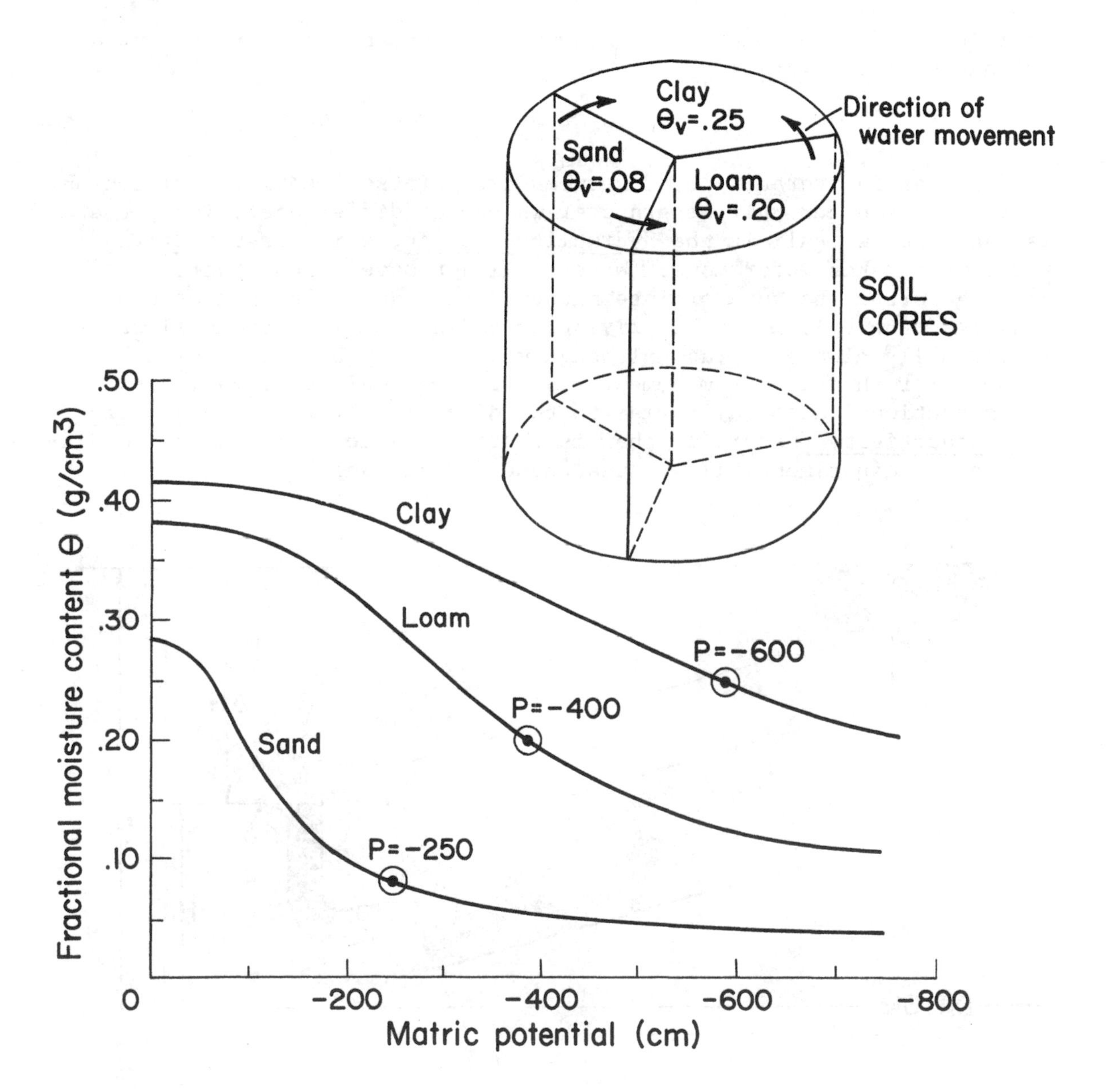

Fig 5-4. The direction of water movement when three soils of unequal moisture content are pressed together is explained by reference to the moisture potential curves of each soil.

G. Soil water movement. When there is a difference in the total moisture potential between two points in the soil mantle, water will move from the highest to the lowest potential. Furthermore, the water particle will always move in the direction of the steepest gradient; that is, not only "downhill" but straight "downhill." But down-gradient may be uphill, as, for example, when a dry soil block is laid on top of a wet one. Soil moisture doesn't always move from wet soil into drier soil either; it can move from drier into wet as it does when gravitational water percolates to groundwater. Fig 5-4 shows how water moves when three unsaturated soils are pressed together. The reason for the movement would remain a mystery if we ignored the moisture potential curves shown in the same figure.

It is customary to combine the gravity (Z) and the pressure potential (P) and call it hydraulic head (H):

$$H = P + Z \qquad 5\text{-}9$$

Differences in hydraulic head between two points in soil is written ΔH. When expressed per unit length over which the difference exists, that is, as ΔH/L, we call it the hydraulic gradient, which is the driving force that makes water move. Water will not move unless there is a driving force. Because of intermittent rain, snowmelt, evaporation and drainage, there is always a driving force in the field and soil water is virtually always moving, although movement may be very slow. To account for the rate of movement requires two additional variables: A cross-sectional area (A) normal to the direction of flow and a hydraulic conductivity factor (K) that is characteristic of the soil. A diagram of an experiment illustrates these definitions.

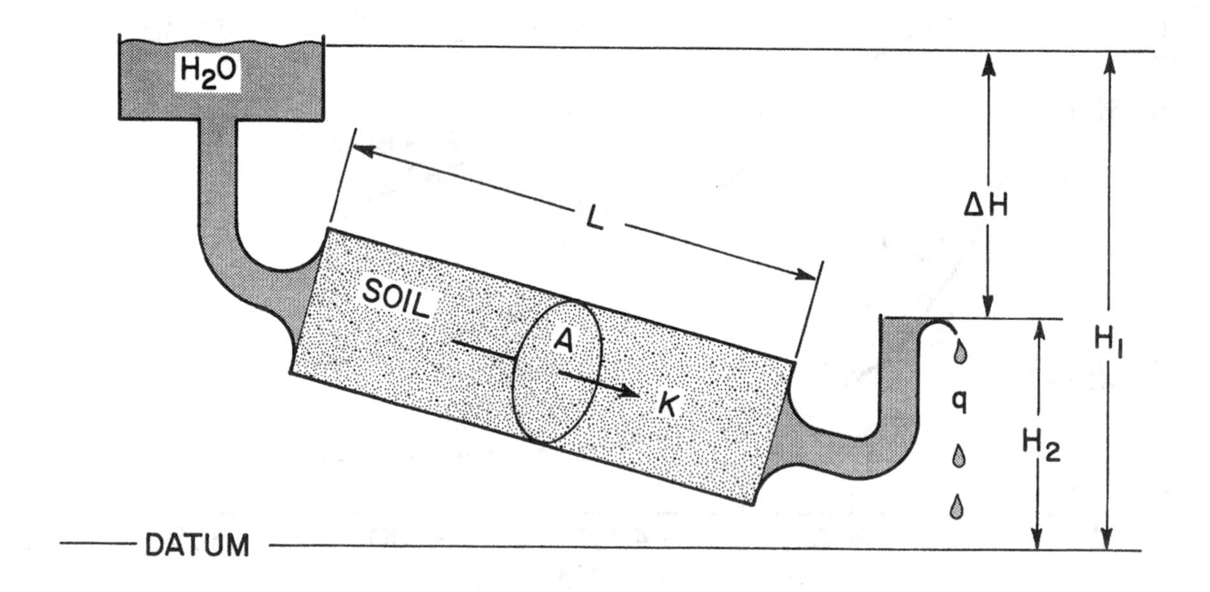

1. Darcy's Law. In 1856, Henry Darcy, trying to solve water filtration problems for the city of Dijon, experimented with columns like the one above and formulated this law:

Water velocity = (Permeability)(Driving force)

$$V = (K)(\frac{\Delta H}{L}) \qquad 5\text{-}10$$

The hydraulic conductivity K has units of velocity (LT^{-1}) because $\Delta H/L$ is dimensionless. When both sides of 5-10 are multiplied by the cross-sectional area A, the volume of flow per unit time (q) is:

$$q = A\ V = A\ K\ \frac{\Delta H}{L} \qquad 5\text{-}11$$

The units of q are L^3T^{-1} and Eq 5-11 applies in the diagram when the soil column is saturated. In practice K is measured on such devices.

> Example: What is the saturated hydraulic conductivity of a soil in the diagram if H_1 is 100 cm, H_2 is 50 cm, L is 40 cm, A is 20 cm^2 and q is measured experimentally as 0.25 cm^3/sec?
>
> Solve 5-11 for K:
>
> $K = (q\ L)/(A\ \Delta H)$
>
> $= (.25cm^3/sec)(40cm)/(20cm^2)(100cm - 50cm)$
>
> $K = 0.01$ cm/sec

Darcy's Law is also used to calculate flow in unsaturated soil, but a complication enters: K is no longer related only to pore size but varies with the soil's moisture content (Θ_v); that is, K is a function of Θ_v as well as a characteristic of the soil. For unsaturated flow, this is written:

$$q = A\ K(\Theta_v)\ \Delta H/L \qquad 5\text{-}12$$

$K(\Theta_v)$, read K-function-theta, may range from 50 cm/day when a soil is wet to 0.001 cm/day as it approaches the PWP. A diagram gives an idea how hydraulic conductivity relates to moisture content:

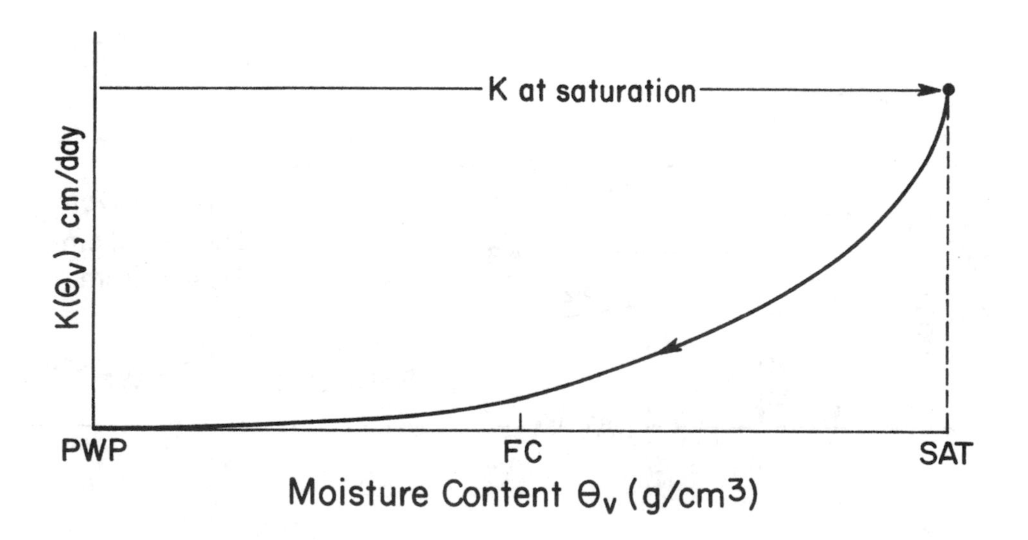

There is hysteresis in this relation also, greatly complicating prediction of soil water movement. However, many useful hydrological and soil science techniques are based on experimentally measured $K(\Theta_v)$ curves.

2. Direction of moisture flow. The student is now prepared to understand the pathways of soil water flow, although its actual computation requires use of large computers. Imagine three points in the soil mantle above the water table in an idealized watershed (see diagram). With tensiometers at Points 1, 2 and 3 we can measure -P as shown, and setting any convenient datum, we can measure Z as shown. The points are 200 cm apart and Point 1 is directly over Point 2 which is just above the water table. Contrary to intuition the pathway of moisture between Point 1 and 2 is not necessarily vertical, but under these conditions will curve down to the right toward Point 3.

At Pt 1: $H_1 = P_1 + Z_1 = -200\text{cm} + 2000\text{cm} = 1800\text{cm}$

At Pt 2: $H_2 = P_2 + Z_2 = -10\text{cm} + 1800\text{cm} = 1790\text{cm}$

At Pt 3: $H_3 = P_3 + Z_3 = -150\text{cm} + 1900\text{cm} = 1750\text{cm}$

The hydraulic gradients are:

From Pt 1 to 2: $$\frac{H_1 - H_2}{L} = \frac{1800\text{cm} - 1790\text{cm}}{200\text{cm}} = 0.05 \text{ cm/cm}$$

From Pt 1 to 3: $$\frac{H_1 - H_3}{L} = \frac{1800\text{cm} - 1750\text{cm}}{200\text{cm}} = 0.25 \text{ cm/cm}$$

The driving force in the direction of Pt 3 is 5 times as strong as towards Pt 2, and the moving water particle bends its path downslope. As water moves under these forces, the potential "field" and moisture contents change, leading water first downward, then slopewise, and later upward to feed evaporation. In the absence of rain, evapotranspiration will gradually lower the hydraulic head in the surface soil, thereby generating strong upward hydraulic gradients ($\Delta H/L$ will be negative).

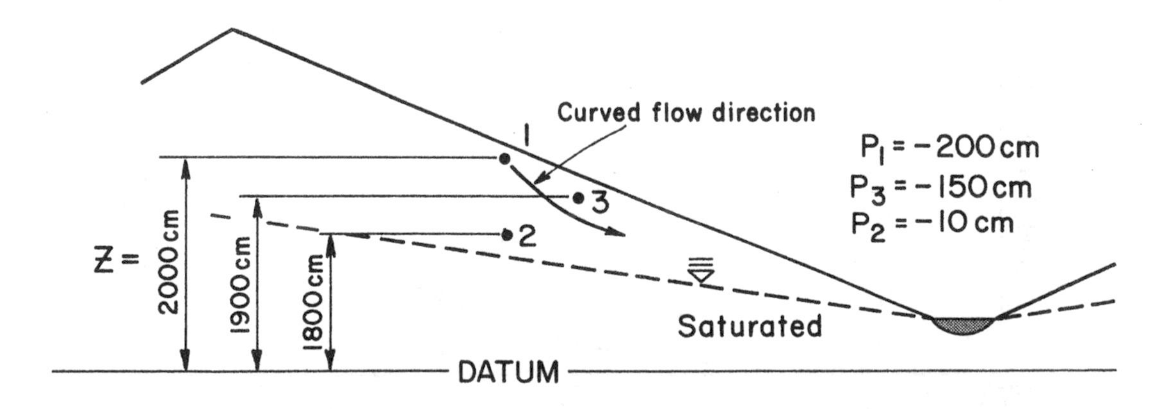

H. Infiltration and percolation. Infiltration is the process by which water passes through the soil surface. Percolation refers to its advance through the soil. Forests generally have a layer of humus on top of the mineral soil and because the depth of humus changes with season, infiltration is defined as the entrance of water into the mineral soil surface. Water moves easily through the humus but its infiltration into mineral soil is a critical process in hydrology. However, under forest cover infiltration is seldom limiting.

1. Factors affecting infiltration and percolation. Water infiltrates into an unsaturated soil in response to matric and gravity potentials. If water is ponded on the surface, the pressure head is an additional potential causing infiltration. The matric potential initially dominates but as water percolates deeper, and soil moisture content increases, the matric potential at the surface becomes progressively less important. When saturation is approached, gravitational forces predominate. The maximum rate at which water can enter a soil under given conditions is termed the infiltration capacity. The rate at which it is actually entering the soil at any time is the infiltration rate. The major factors which affect infiltration rate are:

The quality of infiltrating water (silt, chemicals, debris)
Soil texture and structure (porosity)
Antecedent water content of the soil
Biological activity and organic matter
Depth and type of mulches, forest floor or vegetal cover
Surface soil wettability (some soil is hydrophobic)
Soil frost and ice (concrete or granular)
Entrapped air in underlying soil

The quality of the infiltrating water is often the most limiting. When the soil surface is bare, raindrop energies disperse colloidal materials which tend to plug soil pores and reduce infiltration. Forest litter, humus and fine roots at the soil surface take the energy out of rainfall and allow clean water to infiltrate the mineral soil. Infiltration can fail on roads, trails, and log decking areas due to raindrop impact; the problem is acute on freshly plowed fields and compacted soil of wide extent.

Typical infiltration capacity curves are shown below for an initially dry sand and clay. Zero on the time scale corresponds to the beginning of a storm. It is assumed that rainfall continues to exceed the infiltration capacity for some time, so that no opportunity for recovery of an earlier capacity occurs through drainage into deeper soil.

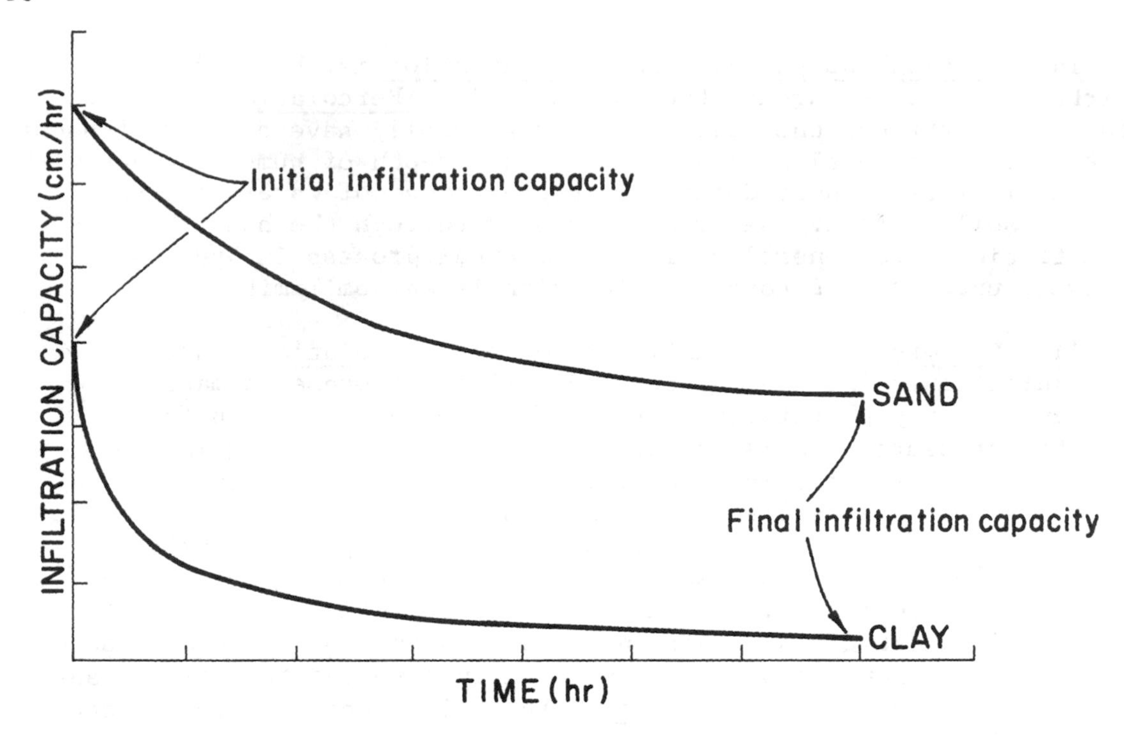

Infiltration capacity decreases with an increase in soil water content until a final infiltration rate is reached. If rainfall ceases or falls below the infiltration capacity, some recovery of the initial capacity will occur. Therefore, actual rates of infiltration will differ, depending on how much drainage time occurs between bursts of rain. Infiltration capacity curves are sometimes used to determine the amount of rainfall that fails to infiltrate, but such estimates are very misleading.

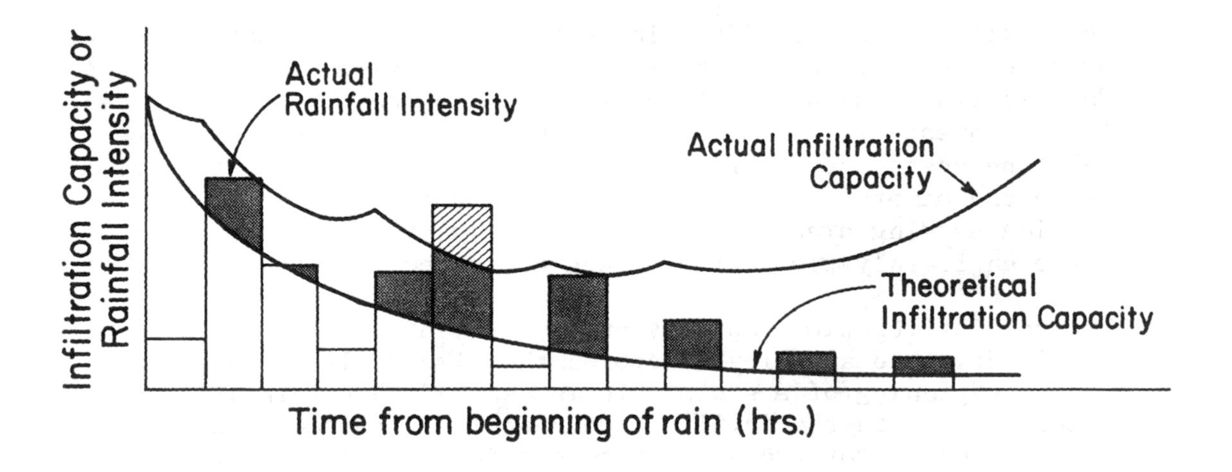

The actual infiltration capacity curve would remain above most of the rainfall intensities and, in this hypothetical case, only the hatched portion would fail to infiltrate.

2. Measurement of infiltration. Infiltration rates measured in the laboratory are of little value to the hydrologist because conditions that prevail in the field at the time of rain entirely control the process. Ring infiltrometers have been used in the field to index the infiltration capacity of different soils, but computation of overland flow from such devices is impossible. So much attention has been paid to the dramatic decrease in infiltration under constant head infiltrometers (where water is ponded continually on the soil surface) that forest managers have been persuaded that any disturbance of the forest floor results in immediate overland flow from the watershed. Actually on most forest soils the disturbance must be severe before overland flow will be generated.

An improvement over the ring infiltrometer, at least for research purposes, is the rainfall simulator which attempts to reproduce the rates and energetics of natural rainfall. The natural dynamics of rain are exceedingly difficult to imitate because neither time nor space variations in rainfall can be reduced in scale. As a consequence infiltration measured on plots almost always underestimates infiltration on basins, and sometimes misleads managers into prescribing infiltration remedies for problems that are not caused by a failure to infiltrate.

The question arises, why not use rainfall and streamflow measurements to estimate infiltration over the basin? This method does not work for reasons that will become clear in Chap 7. Infiltration varies so much from place to place and time to time that its measurement on a watershed basis has been unsuccessful.

3. Percolation and redistribution. During and after infiltration, water is redistributed through the profile. Redistribution is illustrated below, assuming no evaporation or removal by plant roots. Note that infiltration and subsequent redistribution are shown at soil moisture contents less than saturation. Curve 1: Infiltration is rapid and rainfall continues to add to the water

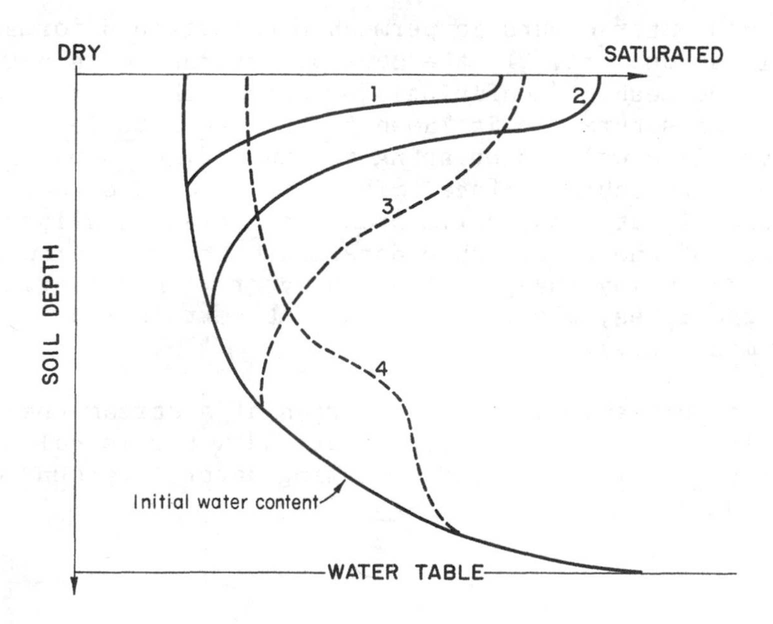

content in the surface layers. Curve 2: The surface is approaching but not necessarily at saturation, while water percolates deeper into the profile. Curve 3: Rain ceases and the water redistributes in a few hours by advancing into deeper soil. Curve 4: In a day or so, gravity and matric potentials draw the water deeper, competing with the surface soils for the supply. The redistribution of soil water will be interrupted by evapotranspiration and rain before equilibrium occurs.

4. Forests and infiltration. Lack of infiltration is a problem on grazed, compacted and cultivated soils but not usually on forest soils. Chiefly because the forest floor absorbs the energy of falling rain and permits clean water to penetrate to the mineral soil layers, the infiltration capacity of forest soils is nearly always greater than prevailing rainfall intensities. To the degree that forest floors are disturbed by cultivation, grazing, repeated burning, logging and road building, infiltration may be impaired and overland flow may occur. It is not the cutting of trees that impairs infiltration but the abuses that follow or accompany tree harvesting. Overpopulation by big game animals, over-use by recreationists, hunters and fishermen, and indiscriminate use of mechanical equipment can cause infiltration problems, but cessation of the abuse usually results in rapid recovery. At some times and in some places recovery is slow, but continued impairment of infiltration capacity in forest and wildlands usually has resulted from a long history of abuse or a combination of several types of abuse.

I. Ground water. Ground water hydrology is a field of study mostly beyond the scope of this text. However, ground water is a major source of usable water in many parts of the world and where forests exist they have an appreciable effect on ground water, particularly when the water table is near the surface.

1. Definitions. Refer to Figure 5-5 and relate the following terms to the diagram:

Ground water occurs in permeable underground formations called aquifers. If the zone of saturation is not beneath an impermeable (confining) formation, the surface of the zone of saturation is known as the water table. The water level in a well penetrating the unconfined aquifer reveals the water table surface. The surface of the zone of saturation is at atmospheric pressure, unless confined. The slope of the water table determines the direction of flow. Direction may change during the year with recharge and discharge rates, particularly in flat terrain (coastal plains, swamps, etc.).

The water table generally emerges at a stream channel or springhead. A stream fed by ground water is called an effluent stream. A stream losing water to ground water is called an influent stream.

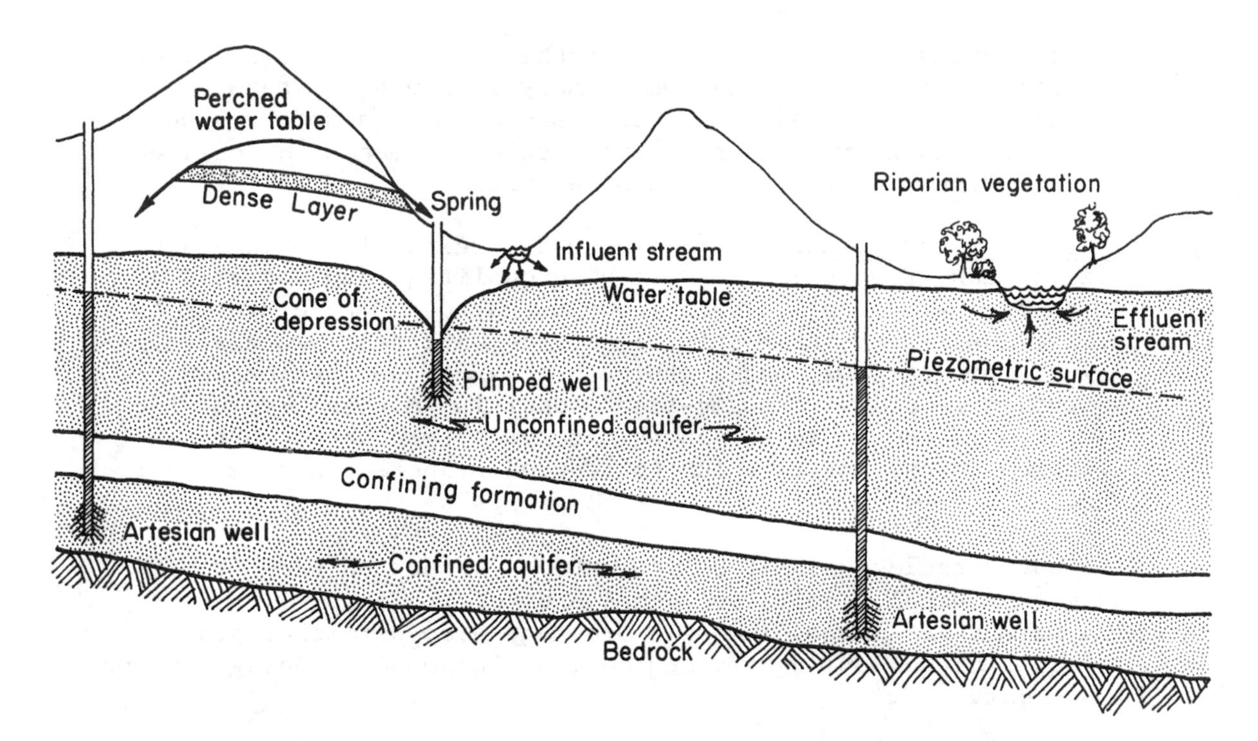

Fig 5-5. Nomenclature and characteristics of ground water. The diagram is for definition of terms, not to illustrate typical ground water conditions.

A perched aquifer is a special case of the unconfined aquifer. It occurs when a relatively impermeable zone such as a clay lens traps water as it percolates toward the main ground water body. Perched aquifers are frequently the source of springs, intermittent streams and some stormflow. Perched water tables fluctuate during wet periods, and may disappear completely during dry periods.

A pumped well in an unconfined aquifer creates a cone of depression. Drawdown of the water table produces a hydraulic gradient in the direction of the well. Permeability determines the rate at which water will move to the pumped well and thus the yield of the well. Wells in the Southeastern Piedmont yield only 5 to 10 gallons per minute (20 to 40 liters/min), while wells in limestone aquifers may yield 2 to 3 thousand gpm (8 to 11 m^3/min).

A confined aquifer, also called an artesian aquifer, occurs where water is confined below a relatively impermeable geologic formation. Water in a confined aquifer is under pressure and as a result the water will rise into wells penetrating the confining formation. An imaginary line projected across the water levels in a series of artesian wells is the piezometric surface. A flowing well results when the piezometric surface is above the ground surface. Recharge to a confined aquifer occurs generally where the confining formation emerges at the surface, often in higher ground many kilometers away. Many confined aquifers are huge and supply all the water used by man in some regions of the country. Forests have little influence on confined aquifers except in the recharge zone.

2. Forests and ground water. Because forest soils maintain high infiltration, forest land favors high quality ground water. At the same time, forests prevent some recharge and take drafts from shallow aquifers by high evapotranspiration rates. In coastal plains, forests lower water tables by evapotranspiration and maintain a relatively "swamp-free" condition during most of the year. It is common in these regions for water tables to rise following forest cutting, due to reduced evaporation of cutover land.

High water tables usually have an adverse affect on tree growth, and growth usually improves when the water table is lowered by a network of drainage ditches. Drainage also permits improved vehicular access during wet seasons, an economic advantage during timber harvest.

Further readings.

Brady, N. C. The Nature and Properties of Soils. 8th Ed. Macmillan Publ. Co., N.Y., 639 pp., 1974.

Freeze, R. A., and J. A. Cherry. Groundwater. Prentice-Hall, Englewood Cliffs, N.J., 604 pp., 1979.

Hillel, D. Soil and Water: Physical Principles and Processes. Academic Press, N.Y., 288 pp., 1971.

Problems:

1. A 25 cm^3 sample from a 2-foot-deep field soil weighs 35.0 g. After oven drying, it weighs 31.0 g. Determine bulk density, water content by volume, by weight, and in inches. If one half inch of water is added to the field, what is the new water content by volume?

2. Can water in unsaturated soil flow into animal burrows, old root cavities or drainage pipes? Why? Under what conditions will water flow into such holes?

3. Compute infiltration rate in cm/sec under conditions below, assuming that saturated hydraulic conductivity of the surface soil is 0.002 cm/sec:

Water surface

↑
1 cm
↓ . Pt 1 Soil surface

↑
1 cm
↓ . Pt 2, matric potential -100 cm

In a few seconds, moisture content at Pt 2 has increased and its matric potential is now only -5 cm. What is the new infiltration rate?

4. Subsurface water flow affects road location and construction. Before construction at Sections a and b below, it is estimated that the average slope-wise velocity of soil water was $\bar{V}_a = 0.1$ cm/hr, and $\bar{V}_b = 1.0$ cm/hr. The saturated hydraulic conductivity K for the soil at both Sections a and b is 15 cm/hr. Compute the total flow of water through Sections a and b before road construction, then after construction. What do you conclude? What do you recommend?

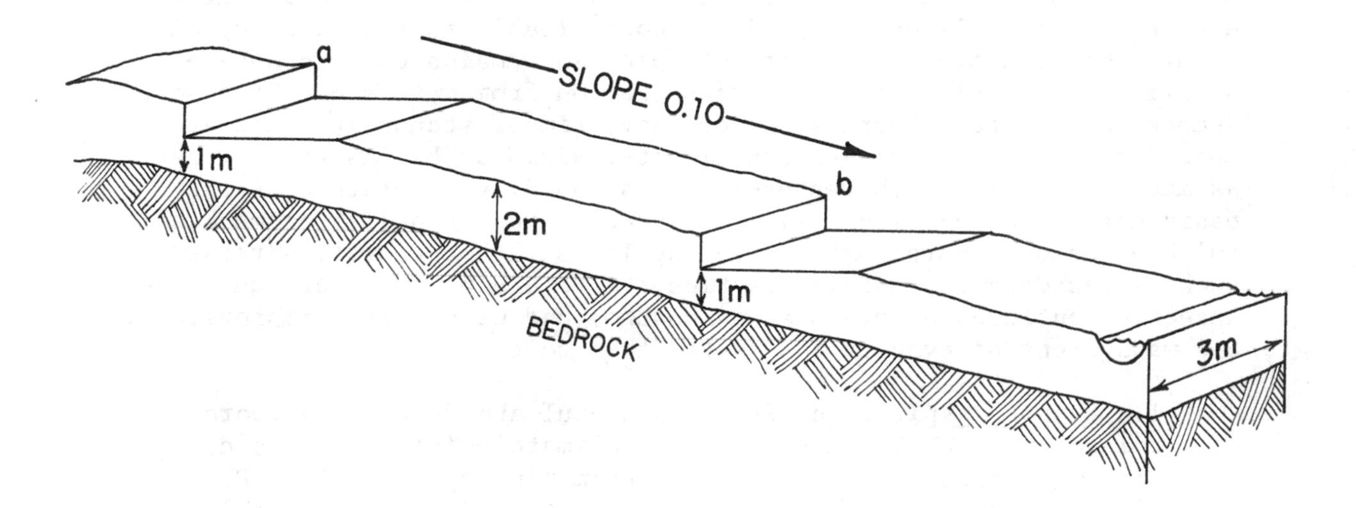

5. Normal streamflow from an eastern U.S. basin is about 1 $m^3/min/km^2$. If bedrock is impermeable and the average soil mantle depth is 4 m, what is the change in soil moisture content by volume produced by one month's normal streamflow from a 100-ha basin? (Disregard recharge by rainfall and loss by evapotranspiration.) How many cm of streamflow did the soil drainage represent?

CHAPTER 6

EVAPORATION AND EVAPOTRANSPIRATION

A. Introduction. More precipitation is disposed of through evaporation than through streamflow and storage combined. In temperate regions about 70% of precipitation evaporates from the land and only 30% goes to streams and soil water storage. Many years of precipitation and streamflow records around the world show that evaporation in a given region varies less from year to year than precipitation and streamflow, that is, evaporation is first served and streamflow is last served from annual precipitation.

Research has shown that evapotranspiration from land and open-water evaporation from lakes and ponds is controllable to some degree, although the economic feasibility of doing so remains open to debate. Surprising reductions in evapotranspiration from watersheds have been demonstrated after clearfelling of heavy timber stands in humid areas; some reductions have caused annual water yield of basins to increase as much as 70 cm, which represents an astounding alteration of the basin energy balance (Eq 2-4). Evaporation reductions of as much as 40% have been measured after treating lakes with evaporation retardants such as hexadecanol, but in practice the retardant films are unstable. Research continues on evaporation control and water yield improvement, but management of evaporation is far from routine.

1. Evapotranspiration (E_t) is a useful and descriptive word coined by C. W. Thornthwaite, the climatologist who devised the first generally used method for estimating regional E_t. Physicists prefer the original word evaporation (E), since the latter includes by definition any process by which liquid water in plant, soil or pond becomes a vapor. Fig 2-1 shows that E_t is the sum of several components:

$$E_t = T + I_t + E_s + E_o \qquad 6\text{-}1$$

where T is transpiration loss by evaporation, I_t is total interception loss, E_s is evaporation from soil, rock and paved surfaces and E_o is open-water evaporation, that is, from streams, ponds, and open-water swamps. In his original work Thornthwaite ignored interception losses and as a consequence the term "evapotranspiration" remains confused in the literature; some authors include interception loss in E_t, some don't. Since E_t is a term intended to include all components of evaporation from land areas, interception loss cannot be left out. Much of the influence forest has on E_t is expressed in the interception component.

2. Potential evapotranspiration (PE_t) is a troublesome term coined by Thornthwaite in an effort to establish a regional index for the maximum possible evaporation under given climatic conditions. Note that this does not mean "maximum possible evaporation." The basic definition of PE_t is full of "ifs": PE_t is the rate of evaporation that would take place per unit area if:

> The plot of vegetation were in the midst of a large, unbroken, similarly vegetated stretch of land.
>
> Soil moisture were so plentiful that uptake by plants would not be inhibited.

The rate of evaporation from a plot in the middle of the Amazon jungle, where vegetal cover is complete and it rains to replenish soil moisture almost daily, might serve as an example where actual evapotranspiration (AE_t) would equal PE_t all the time. Another example might be a hayfield in the midst of a very large irrigation district, all of which is under an irrigation schedule that prevents soil moisture from falling below field capacity. An important consequence of this definition is that under the two conditions specified E_t ought to be entirely under meteorological control. The simplest index of atmospheric heat loading is the mean monthly air temperature reported from standard weather stations throughout most of the world. Using such data, Thornthwaite generated regional maps of PE_t.

> In theory AE_t should never exceed PE_t, since by definition PE_t is maximum for the climatic conditions prevailing. In practice however, it is often found that actual evaporation exceeds estimated PE_t, as for example, measured interception losses from forest in winter. Any effort to estimate PE_t is method-bound; that is, a particular formula must be used to calculate a value, since there is no device big enough to measure PE_t under the two conditions specified in its definition. Since PE_t is an index computed by a particular method, the student should be aware of the assumptions made in developing the method. Thornthwaite's PE_t, for example, is based solely on mean air temperature and an estimate of possible sunshine duration.

3. Transpiration (T) is the vaporization of water from the living cells of plant tissues (excludes interception loss). Leaves and twigs are covered with non-living layers of cuticle or suberized tissue which are largely impervious to vapor transport; "living cells" are within plant surfaces, behind the stomata (the "breathing" pores of leaves) or lenticels (occasional pores in the suberized bark of twigs). Transpiration is a physical process but its rate is under some stomatal control, the only component of E_t under any degree of physiological control.

4. Total interception loss (I_t) is the vaporization of water intercepted during precipitation (rain or snow) from living or dead plant surfaces, including leaves, twigs, stems, down trees, forest litter and humus layers. Effort to measure interception loss has resulted in identification of two sub-components: Crown interception loss (I_c) from leaf, twig and stem, and forest floor interception loss (I_f). Interception loss has no opportunity to become soil moisture, and therefore no opportunity to transpire or drain to streams. I_t is largely determined by the size, structure and above-ground arrangement of plants, as well as by the total surface area of plant materials.

5. Soil evaporation (E_s) is the vaporization of water directly from the mineral soil surface. E_s is minor under full forest cover because the forest floor insulates the mineral surface from radiation and prevents air movement over it. E_s increases as the vegetal material imposed between soil and sky is reduced, and replaces E_t in part when vegetal cover is absent.

6. Open-water evaporation (E_o) is vaporization from water surfaces free of vegetal cover, either emergent or overhanging. On a smooth, quiet water surface, the rate of water loss will be related to the temperature of the surface layer and the water vapor pressure of the air immediately above. The temperature of the water determines the vapor pressure at its surface, and the rate of evaporation is proportional to the vapor pressure difference between the water surface and the air. Evaporation will cool the water surface, reducing the driving force, unless there is an outside source of energy to maintain the difference in vapor pressure. Convection currents, water depth, waves, wind, salinity and surface contaminants all affect E_o, but only three factors need be included in the major causes of open-water evaporation. Dalton (about 1800) proposed an equation for predicting E_o that usually takes this form:

$$E_o = C\,u\,(e_o - e_a) \qquad 6\text{-}2$$

where u is wind speed measured above the water surface, e_o is the vapor pressure of the surface water (a direct function of water temperature), e_a is the vapor pressure in the air above the water and C is a coefficient to take care of unit conversions. The formula states that if we know the temperature of the surface water, the vapor pressure in the air (from psychrometry) and the wind speed over the water (from an anemometer), E_o can be calculated. Unlike vegetated land, water is usually a simple, flat surface, so refinements of Eq 6-2 give fairly accurate estimates of E_o.

Lake Ontario under a cool, humid climate evaporates 50 cm of water per year, while Lake Meade in a southern California desert evaporates 216 cm per year. Evaporation from stored water represents a severe economic loss in arid and semiarid areas, both from a water quantity and quality viewpoint.

Evaporation pans are metal containers 1.22 m in diameter and 30 cm deep which rest on a slatted base just above the ground surface, devised to give an index of evaporative conditions. They are read daily at Class A Weather Stations. The pan water temperature seeks that of the air but is never the same, so the index is a direct measure of neither E_o nor E_t. Conversion coefficients ranging from 0.5 to 1.0 are used to approximate land and water evaporation from pan evaporation but these are unreliable. Even if floated in lakes, the pan water will assume a temperature different from the lake water and e_o (Eq 6-2) in the pan will not be the same as e_o in the lake (pan water will usually be warmer during the day and cooler at night).

While theoretical formulae for E_o work fairly well over water, similar formulae derived for E_t over vegetated surfaces are not refined to the point of reliability. The complexity of vegetal surfaces as compared with open water defeats practical efforts to compute transpiration, interception loss and soil evaporation from weather data alone. Primary information on the effects of vegetal cover has come from experiments on catchments, supplemented by measurement of soil moisture, radiation, wind speeds, vapor pressures and temperatures in and around crops and forest stands.

Under full forest, E_o and E_s may generally be neglected and E_t becomes the sum of T and I_t. If vegetation is eliminated then E_t is entirely soil evaporation, as for example, in deserts. Man's management of non-arid land may reduce or increase E_t, by shifting vapor losses between the four components in Eq 6-1.

B. Evaporation as a process. Chaps 4 and 5 introduced the energy status of water. Water is a compound with a molecule containing one oxygen and two hydrogen atoms. These atoms are held together in a rigid position by chemical bonds. The hydrogen atoms, however, have some bonding strength, generally called hydrogen bonds, left over for alignment with other molecules. In addition, all molecules attract each other by weak forces called van der Waal's forces. In ice, most of the molecules of H_2O are rigidly arranged in a lattice structure. As the temperature rises due to an input of energy, the molecules vibrate more actively, loosen the hydrogen bonds, and the ice melts. As the temperature and molecular activity increase further, more and more molecules escape from the water surface (some condense back into the water as well). A constant supply of heat energy is required to keep the escape (net evaporation) going. Net evaporation will soon stop without an energy source.

1. Conditions necessary to sustain evaporation. Two general conditions sustain evaporation: Energy and water availability.

Availability of energy. Some of the short-wave energy from the sun is converted into heat in the leaf, water body or soil. Some heat warms the water and increases evaporation rates, while some warms the air or other materials nearby.

The heat used to warm materials without a change in state is called sensible heat ($H + G$ in Chap 2). When warm air drifts from one area to another, it transfers energy (H); thus solar energy developed into sensible heat in one area may be available to evaporate water in another (advective energy). High evaporation from desert oases illustrates advective sources of energy.

Topography and the elevation of the sun in the sky affect energy availability in complex ways. Fig 6-1 shows how the angle of the sun's rays interact with topography to provide different solar loads on catchments oriented south and north. While summer loading of short wave radiation is similar, winter loading on the south basin is over 3 times that on the north. Diffusion of radiation by clouds and atmospheric dust tend to reduce the difference, so it is not known exactly how the energy available for E_t may be affected by aspect.

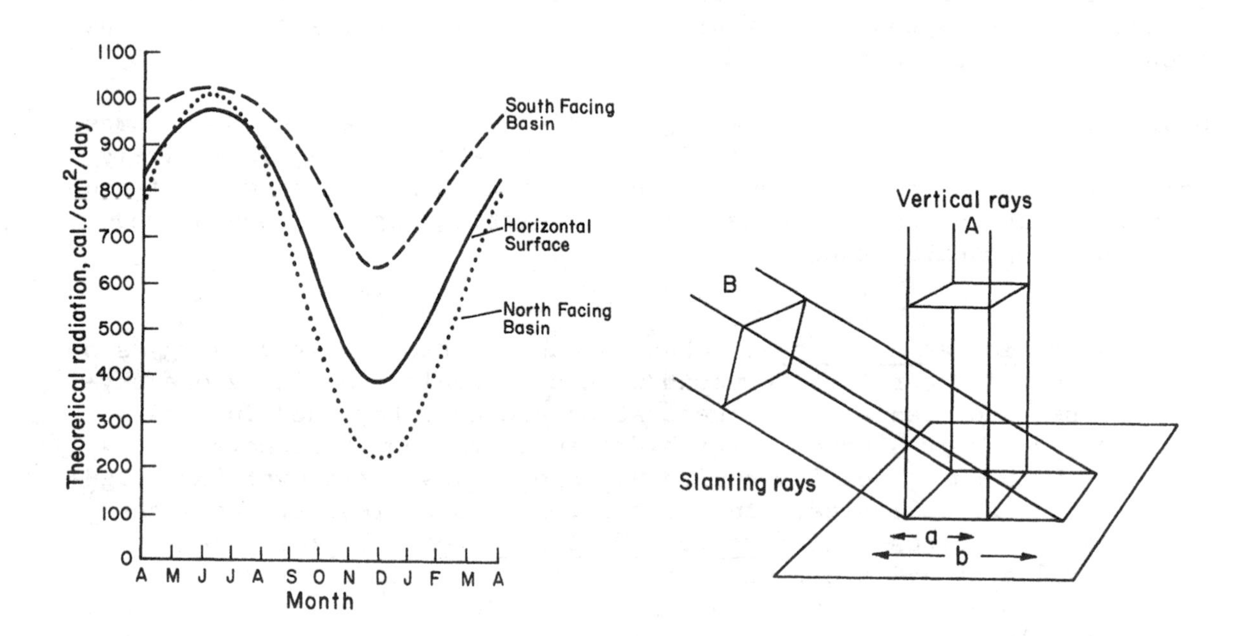

Fig 6-1. The effect of solar angle on total daily radiation (short-wave only) received by two mountain basins, as compared with the receipt by a horizontal surface. The two basins lie on the north and south slopes of an east-west ridge in the southern Appalachians (after Swift 1967).

Part of the sun's radiant energy is converted into the mechanical energy that drives the wind, which in turn produces turbulent mixing of the air and water vapor near the ground. Turbulent mixing increases the efficiency with which available energy is applied to evaporate water from land areas. In effect, it keeps the air near the ground from becoming saturated, and so maintains a higher vapor pressure gradient at the evaporating surface.

Availability of water applies not only to the amount of water present but also to its availability for evaporation. An analogy: A sealed bottle of water may be warmed by many calories but little evaporation occurs. If the cap is removed, evaporation begins but not all the added heat will evaporate water because only a small surface is exposed and the vapor must be transported through the mouth of the bottle. If the water is poured out over a smooth level surface, the added heat will evaporate water more rapidly. Finally, if the surface is rough instead of smooth, a very efficient evaporating body is formed and added heat will be converted readily to latent heat of vaporization. In a general way, this simple analogy applies respectively to water within the soil, water within the plant, and water on the surface of plants. Soil moisture held between field capacity and the wilting percentage is often erroneously thought to be equally "available"; in fact there are always degrees of availability.

Consider the interactions between heat and water availability as each affects the evapotranspiration rate. There is little or no evaporation in a hot desert, although there is plenty of energy available. The energy is reflected or heats the air, soil, or is advective to other areas. On the other hand, there is little evaporation (or sublimation) in Antarctica, where there is plenty of water, but little available heat because of the low angle of the sun and the high albedo of snow. In between there are many degrees of relative availability.

C. Evaporation from bare soil. About one-third of the contiguous United States is classed as arid or semi-arid. Vegetal cover in these areas is usually less than 25% and rainfall is less than 30 cm per year; streamflow averages less than 3 cm per year. Most of the remainder evaporates from or through the surface of soil or rocks and is lost to water supply. Recovery is made in some dry areas through water harvesting, which simply means paving or sealing a small catchment or plot to force rainfall immediately into cisterns, stock-watering tanks, or ponds. The ancient peoples of the Negev Desert were skillful at water harvesting.

1. Process of soil evaporation. Through many years during which E_t was nearly always greater than P_g, the soils of arid lands dried to depths of 15 m or more. This proves that, even in the absence of vegetation, soil moisture can slowly return to the surface and evaporate. Two modes of travel are known.

Film flow of water occurs under matric potential (-P) gradients to the zone where evaporation takes place. Experiments in the laboratory show that movement becomes negligible as the moisture content reduces toward the permanent wilting percentage. From these experiments, it was once thought that soil moisture a few centimeters below the soil surface could not evaporate and that even quite arid lands could be kept in fallow until enough soil moisture accumulated to grow a crop.

Vapor flow of water occurs from the zone where evaporation takes place in the soil to and through the surface. Temperature and vapor pressure gradients move the water toward the surface. While very slow, vapor flow can account for the drying of desert soils to great depths. It has been suggested that such movement may be aided by barometric "pumping," caused by diurnal or irregular changes in barometric pressure. Dry air is forced downward into the soil when the barometer rises and soil air escapes as it falls.

That temperature plays an important role in both modes of flow is evident in the experience of highway engineers who long ago noticed the winter migration of soil water to the underside of road pavements.

D. Evaporation of intercepted water. In nearly all cases, water caught and held by vegetation is evaporated from the surface. In a few cases it is thought that some of the intercepted water is absorbed by the tissues of plants to join the transpiration stream, but evidence indicates that absorption of intercepted water is hydrologically negligible. Therefore we must treat interception loss as a component in itself, not just an aspect of transpiration.

Interception denotes a process in which precipitation of any type strikes vegetal material above the mineral soil surface.

Interception storage is the amount of water or snow held by living and dead plant material at any given time.

Crown interception loss (I_c) is the amount of water evaporated or sublimated directly from water or snow intercepted by the crowns of vegetation.

Throughfall (P_t) is that portion of gross precipitation which falls or drips through the crowns.

Stemflow (P_s) is that portion of the intercepted water which collects and runs down stems.

Forest floor interception loss (I_f) is the amount of water caught and lost from the forest floor (L, F and H layers collectively) before it can infiltrate mineral soil.

Total interception loss (I_t) is the amount of water evaporated or sublimated from rain or snow caught by living and dead plant material.

Each time precipitation occurs over forests (50 to 120 times per year in the East), a certain amount never reaches the mineral soil surface and never contributes to soil moisture, ground water or streamflow. Most of a 5 mm rainstorm will be lost by evaporation from the crowns and litter; not much more will be lost from a 100 mm storm but obviously the percentage loss from the larger storm will be far less. Total loss will depend largely on the above-ground mass of plant cover and the number and length of drying periods between storms. A drying period of 12 hours is usually required to separate successive rainstorms for interception computations.

If we assume 2 mm interception loss from each of 60 storms during a year, the total interception would be 120 mm. Assumptions are not good enough, however; interception losses must be measured in different regions, climates, and forest types. In spruce forests in England, where rainstorms are frequent and small, interception losses as high as 38% of annual precipitation have been measured.

Methods for dealing with interception of snow have not been developed because of the peculiar physics and erratic behavior of intercepted snow. There is lack of agreement on how important snow interception loss is to water supplies.

1. The availability of water in the interception process is controlled by two sets of factors:

   Vegetative factors

   The total surface area of the living and dead plant material (percent crown coverage, tons of forest floor, leaf area index) and the seasonal differences in leaf surface area (deciduous or non-deciduous).

   The nature of the surfaces (roughness, smoothness, wettability, litter absorptiveness).

   The arrangement of leaves and stems (height of plants, needle fascicles, amount of dead twigs and limbs, shingling effect of leaves, type of branching.

   Weather factors

   The number and spacing of precipitation events.

   The intensity of rain or quality of snow (determines the percentage of gross precipitation penetrating crowns and forest floor).

   Wind speeds during and following precipitation (shaking and exposure of stems and leaves).

Water on the surface of plants is more readily available for evaporation than any other water in the drainage basin. Consequently, when leaves are wet, interception loss proceeds at a rate up to 5 times transpiration from surface-dry vegetation.

2. <u>The availability of energy</u> to evaporate intercepted water is the key to understanding these rapid rates of loss. The sources of energy are net radiation (R_n) and sensible heat (H) moving into the forest from upwind areas. The latter depends on air speed and temperature and the former depends on solar radiation after all the reflectance and re-emission components of radiative energy are accounted for (Chap 2). Through turbulent mixing, air speed plays an additional role by increasing the efficiency with which energy is applied to evaporating surfaces.

Windspeed and radiation at the forest floor are particularly effective in hardwood forest floor interception loss (I_f), which reaches its maximum rate not in summer, as one might expect, but in late winter and early spring when trees are leafless and air speeds near the ground are greatest. This accounts in large measure for the spring and fall forest fire hazard in hardwood forests of eastern U.S. After brief summer rains, stored (sensible) heat in the stems, leaves and ground supply energy to evaporate small amounts of intercepted water, perhaps up to 0.25 mm per rainstorm.

3. <u>The effect of interception on transpiration</u> has been much debated.

> Does transpiration continue while intercepted water evaporates? Until recently it was thought that the rate of transpiration would be greatly reduced, if not altogether eliminated, while water on the surface of the leaf evaporates. However, the upper portions of plant crowns may be surface-dry and transpiring while the lower or inner portions are still wet with intercepted water. Some transpiration may continue while intercepted water evaporates, although the sum ($T + I_t$) cannot, of course, exceed the energy supply available at the plant surfaces.
>
> Over a day or a week during which some rain falls, is the total vegetal evaporation ($T + I_t$) the same as it would be if that rain were added directly to mineral soil under the vegetation? In other words, with that water in the soil, would transpiration proceed as fast as interception loss? Far from an academic question, the answer has much to do with the role forests play in evapotranspiration from watersheds. Research has shown that intercepted water evaporates much faster than transpired water, and therefore much of the interception loss represents an additional loss in the water balance of watersheds.

4. <u>Measurement of interception loss</u>. Note this obvious relation.

$$I_t = I_c + I_f$$

Forest floor interception loss (I_f) is measured by collecting and weighing samples of the forest floor. The first sample is collected shortly after rainfall and the average loss of weight in later samples is the amount of intercepted water evaporated.

Forest floor interception averages about 5% of annual precipitation under mature pine and 3% under mature hardwoods, the difference due to the more rapid decomposition of hardwood litter.

Fig 6-2 shows how crown interception loss is estimated in field experiments. I_c is equal to gross precipitation minus the sum of throughfall and stemflow:

$$I_c = P_g - (P_t + P_s) \qquad 6\text{-}3$$

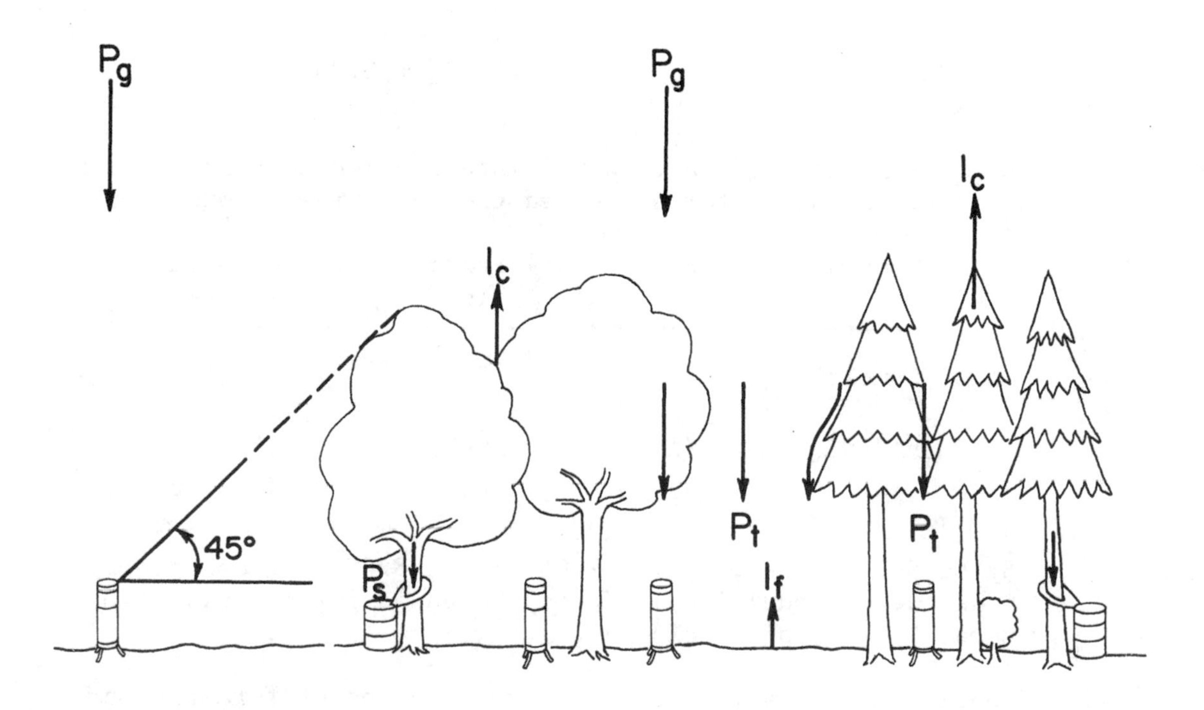

Fig 6-2. Measurement of crown interception loss (I_c) and forest floor interception loss (I_f). About 10 throughfall gages are required to equal the mean accuracy of one gross precipitation gage in the open. I_f is measured by successive weighing of samples of the forest floor.

P_t is caught in rain gages under the canopy. Collars around stems collect P_s and gages in nearby openings estimate P_g. I_c is estimated by storm and by season. P_s is usually only 1 to 2% of P_g, except in beech forests where it may be 5 to 8%. P_s in spruce and fir is hydrologically negligible.

The following equations (Helvey 1971) may be used in temperate regions to estimate annual crown interception loss from annual precipitation (P_g) and the number of storms (n) larger than a trace amount. Mature forest is assumed; a correction for thinned stands may be made by reducing the estimate of I_c proportional to the reduction in basal area, crown cover or biomass.

Pines $\quad I_c = 0.10\ P_g + 0.10\ n \quad$ 6-4

Spruce, fir, hemlock $\quad I_c = 0.21\ P_g + 0.13\ n \quad$ 6-5

Deciduous hardwoods

Summer $\quad I_c = 0.06\ P_g + 0.10\ n \quad$ 6-6

Winter $\quad I_c = 0.03\ P_g + 0.05\ n \quad$ 6-7

Example: Calculate the annual interception loss in cm from a mature hardwood forest, given the following information:

Season	P_g (cm)	# Storms (n)	Computed I_c (cm)
Summer	76	38	8.36
Winter	71	43	4.28
Year	147 cm	81	12.64 cm

By Eq 6-6 and 6-7, I_c is 12.64 cm, or 8.6% of gross precipitation. An additional 3% may be added for I_f, giving a total loss I_t of 11.6% (17 cm).

Interception losses vary both with age and type of forest stand. For example, I_t measured as a percent of P_g in the Southern Appalachians was 12, 15, 19 and 26%, respectively, for mature hardwood, 10-yr-old white pine, 35-yr-old white pine and 60-yr-old white pine. Interception loss accounts for much of the difference in consumptive use of water among forest types.

E. Transpiration. Fig 6-3 shows the pathways and tissues involved in absorption of soil moisture, translocation of the absorbed water through the network of conducting cells and vessels called the xylem, and finally the transpiration (evaporation) of vapor through the stomata. When the stomata are closed, resistance to transpiration is greatly increased, but it is wrong to conclude that the tree operates the transpiration stream by expenditure of its own metabolic energy. Most authorities agree that transpiration is a passive process, not an

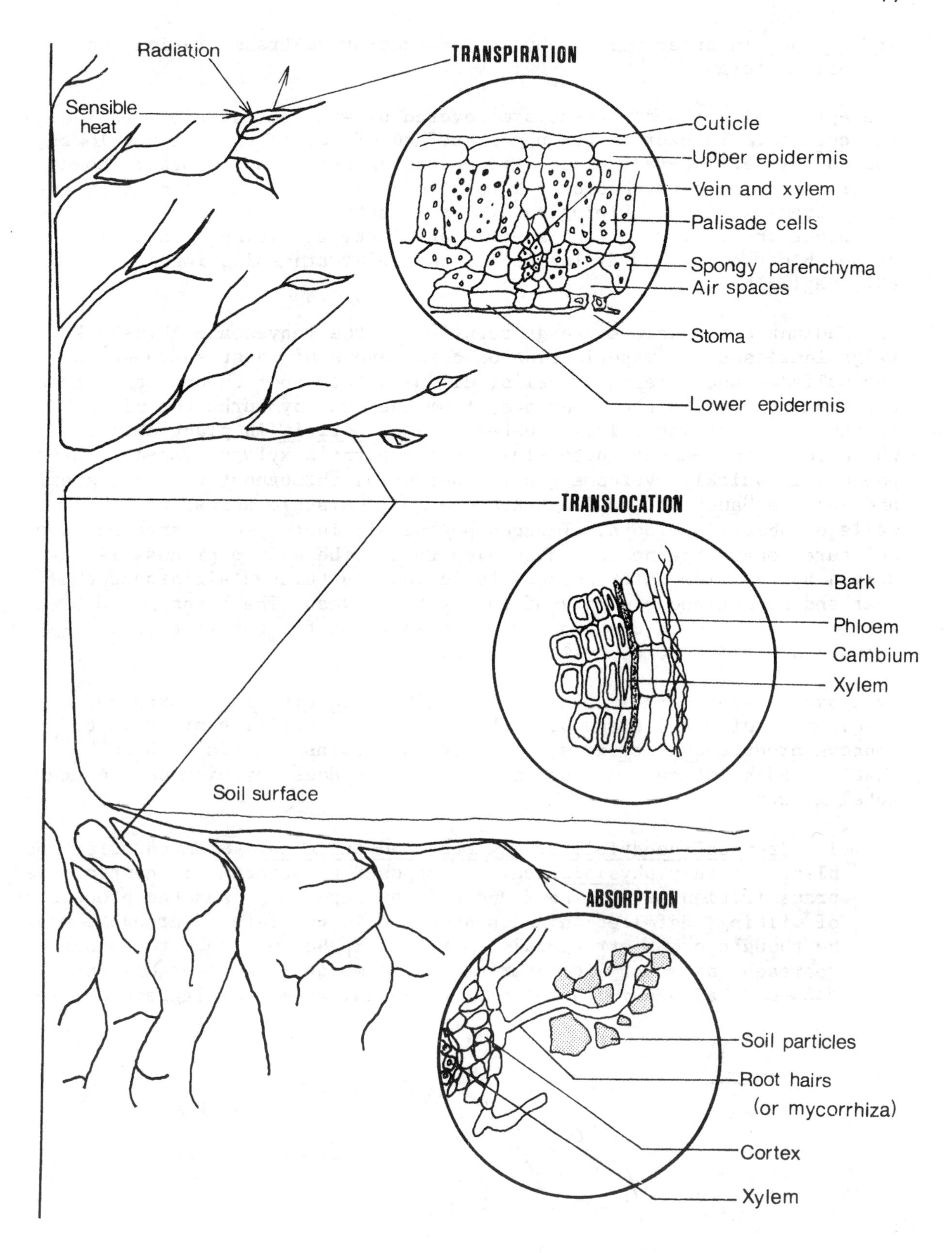

Fig 6-3. The movement of water into and through vegetation. Many trees do not have root hairs but fine rootlets or mycorrhizae serve as efficient absorption surfaces.

active one; in other words, physical forces drive transpiration, not metabolic energy.

The epidermal cells of leaves are covered by waxy cutin layers called the cuticle. Thicker on the upper surface of leaves than on the lower, the cuticle effectively blocks transpiration except by stomata. Some experiments indicate that only 2 to 3% of total transpiration is cuticular transpiration. Biologically, it is sufficient in drought periods to wither the leaves of many species of plant, resulting in abscission or death of leaves. Hydrologically, cuticular transpiration is probably negligible.

The radiant and sensible energy supplied to the leaves and thin-barked twigs increases the vaporization of thin layers of water surrounding the palisade and parenchyma cells, diffuses the vapor through the stoma, and finally carries the vapor away from the leaf by turbulent mixing of the air. The water loss creates a water deficit in plant cells, which leads to negative potentials in the plant's xylary system. These potentials quickly overcome gravity potential throughout the tree stem, and apply a "suction" gradient to move soil moisture across the cell walls of absorbing roots. Induced potentials in the soil serve to move moisture toward the nearest absorbing root. The entire process is driven by very low water potentials in the unsaturated air around the leaf and a continuous supply of energy to leaves. The plant would be entirely helpless in the process if it were not for the fact that stomata are under some physiological control.

Even small water deficits in leaves inhibit quantitative growth and development of plant tissues, although periodic deficits are said to improve product quality in some fruits and grains. It is doubtful that periodic moisture stress in forest trees does anything but reduce total growth.

1. Stomatal behavior, arrangement and function. Stomata offer the plant its only physiological (as opposed to structural) control over transpiration or E_t. Leaf and twig arrangements, and the processes of wilting, defoliation and senescence in the face of drought, may be thought of as structural control of E_t because such responses represent no metabolic interference to maintain a favorable internal water balance in the plant. A typical stoma is diagrammed here:

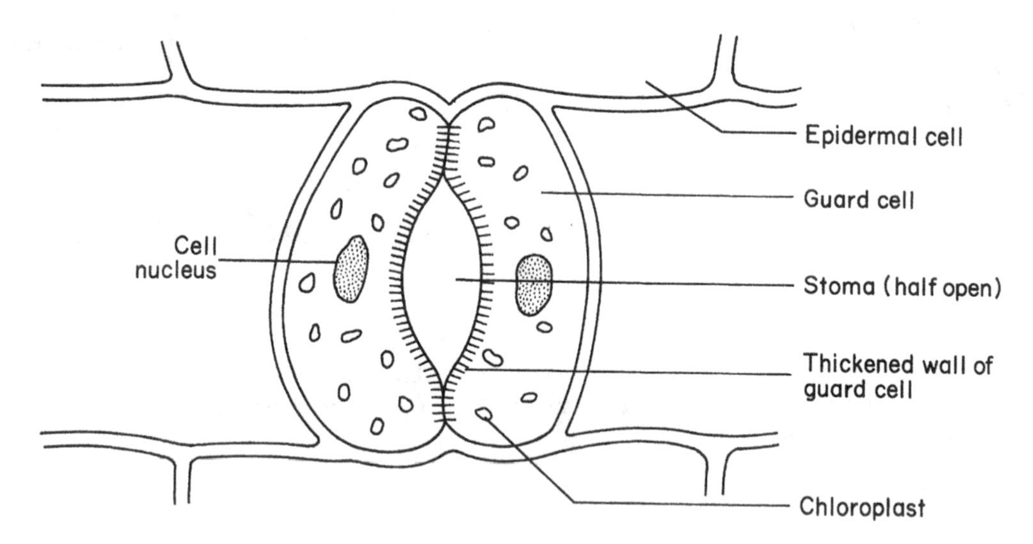

Most trees have stomata only on the underside of leaves. There are from 10,000 (Cornus spp.) to over 100,000 (Quercus spp.) stomata per cm^2 of leaf surface, depending on species, leaf-type and other factors. The following factors, interrelated in complex, poorly understood ways, are known to affect stomatal opening and closing:

Water deficits in the guard cells.
CO_2 concentration and pH inside leaf cells.
Certain qualities and intensities of light.
Temperature of the leaf (thermoperiodism).
Time of day (photoperiodism).
Certain chemicals; for example, phenolmercuric acetate.
Mechanical stimulation of leaves by wind.

The responsiveness of stomata to these factors varies among species. Since transpiration creates harmful water deficits in leaves, why did plants evolve stomata to allow escape of vital water? The answer seems to be that without stomata the inward diffusion of CO_2 would be too slow to sustain competitive growth rates. With stomata, there is no way to prevent excessive loss of water (the H_2O molecule is smaller than CO_2). Transpiration has therefore been called "a necessary evil." However, it cannot be denied that plants grow best when and where transpiration is greatest because that is when and where water is most readily available to sustain both leaf turgor and vapor loss.

2. Water potential in plants. Water moves in plants for the same reasons that water moves in soils (Chap 5). The basic rule is that water moves down the steepest gradient in total water potentials, regardless of the direction of the resultant forces. In plants the direction is up toward transpiring surfaces; movement of water from root into soil (reverse transpiration) has been hypothesized but never observed.

Soil particles and plant cells differ in regard to the meaning of the osmotic and matric-pressure potentials in Eq 5-9 from Chap 5:

$$\Psi = O + P + Z \qquad \text{UNITS: } L$$

For purposes of this outline, osmotic potential (O) is defined as the solute potential of the cell sap, which is the solution within the differentially-permeable membrane of the cell protoplast. This membrane is permeable to water but not to the solute; water enters the cell until the protoplast exerts turgor pressure against the outer cell wall (a wall of cellulose). Under full turgor the cellulose wall exerts a positive potential on the protoplast exactly equal to the negative osmotic potential inside. Although the cell still has affinity for water due to its osmotic potential, it is prevented from absorbing any more water by the pressure of the wall. Under turgor the cell wall will have free water in its cellulose microfibrils, water which can evaporate into spaces within the leaf or can be passed to other cells which are not equally turgid. Cell wall potential is positive when the cell is turgid and negative when the cell is plasmolyzed:

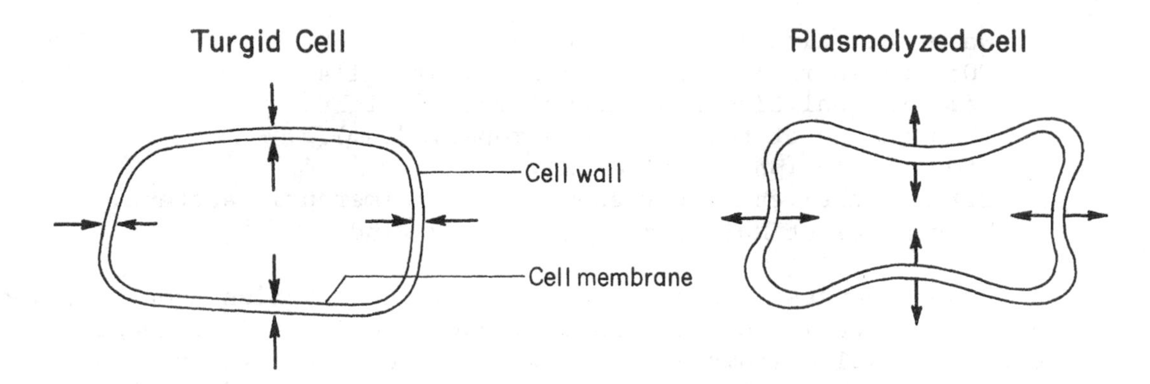

The matric-pressure potential (± P) in Eq 5-9 is the sum of wall pressure, imbibitional forces in the cellulose microfibrils and capillary forces between cells. These can be separated from osmotic potentials of the cell sap only by highly specialized techniques, therefore we will combine (0 + P) in Eq 5-9 into plant water potential Ψp:

$$\Psi = \Psi_p + Z \qquad \text{UNITS: L} \qquad 6\text{-}8$$

The unit is still energy per unit weight (Chap 5). Plant water potential in transpiring trees is nearly always more negative than soil matric potential at the roots, giving the tree "suction" energy to absorb water. The xylem cells and vessels, which are dead, serve to pass these potentials from living cells of the leaf to living cells of the roots, thus completing the transpiration stream. When the tree dies, the transpiration stream is permanantly destroyed.

The gravity potential Z may be measured to a convenient datum, just as in soil. The only simple device to estimate Ψ_p in situ is called the Scholander pressure bomb, after its inventor. A twig is cut quickly from a plant and inserted in a high-pressure chamber with the cut surface exposed through a small hole in a pressure seal. Gas is pumped into the chamber until a wet surface appears over the cut stem. The positive potential indicated by a gauge in the chamber is taken to be numerically equivalent to the negative plant water potential that existed in the plant tissue. These values range from -5 m to -300 m, depending on soil moisture content, vapor pressure deficits between plant and air, and position within the plant.

3. Water movement in plants. The flow of water from soil to air through plants is often explained by analogy to Ohm's Law which states that the flow of current in a wire is directly proportional to voltage (potential) and inversely proportional to resistance. Since resistance (r) is the reciprocal of conductivity (K in Eq 5-12), the flow (q) in plants is:

$$q = A \frac{1}{r} \frac{\Delta\Psi}{L} \qquad 6\text{-}9$$

But resistances along the soil-plant pathway are complex. If the soil is moist, resistances at the root surface and at the stoma are said to be the strongest, whereas root and stem (xylary) resistances are relatively weak. The total resistance r is the sum of the resistances in each pathway (as in complex electrical circuits), so that the hydraulic conductivity K of the forest transpiration stream may be represented:

$$K = 1/r = 1/r_{soil} + 1/r_{root} + 1/r_{stem} + 1/r_{stoma} + 1/r_{air} \qquad 6\text{-}10$$

Subscripts indicate each pathway in the transpiration stream. Plant-soil resistances change minute by minute and are almost impossible to compute. Eqs 6-9 and 6-10 help explain the flow of water through plants but the only absolute way to measure transpiration is to weigh the periodic loss of water from a sealed potted plant or a weighing lysimeter. The latter is a large "pot" or container suspended in such a way that hourly loss of weight can be recorded. Used to measure crop water loss, even the largest lysimeters (about 60 metric tons) are too small to grow a forest tree, let alone a stand of trees. The best estimate of in situ transpiration from forest stands is deduced from catchment experiments, coupled with interception and soil moisture measurements.

Example: Eq 6-9 provides insight into water absorption by depth under forest trees. Suppose a transpiring tree has its roots distributed in soil according to this diagram. The mean distance between randomly located soil moisture "particles" and the nearest absorbing rootlet varies from about 1 cm at 20-cm depth to 10 cm at 250-cm depth:

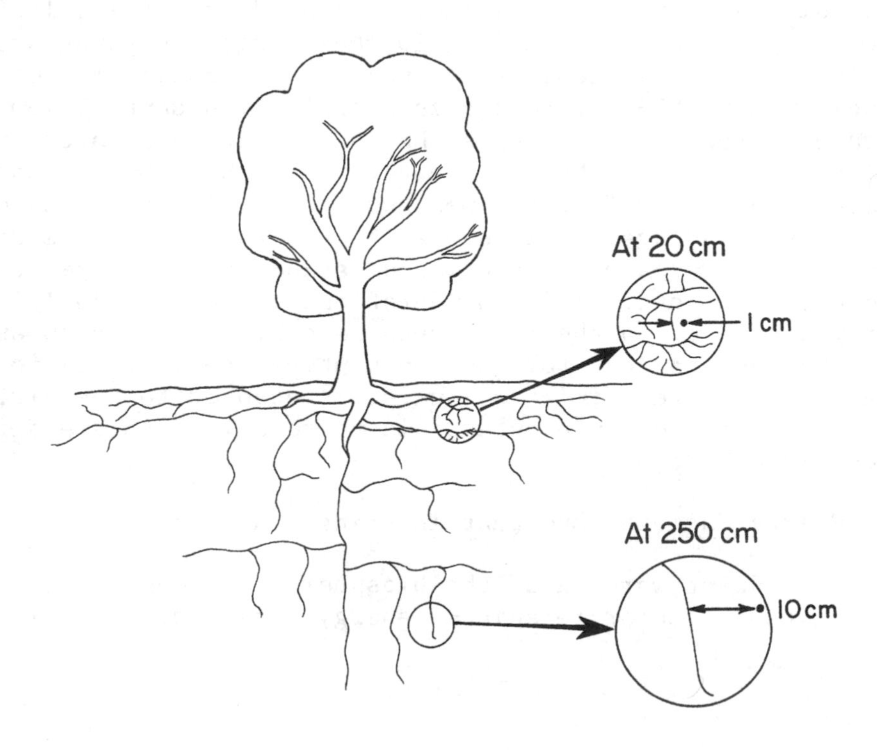

Furthermore, assume the following:

	At 20 cm	At 250 cm
Surface of roots/volume of soil	120 cm^2/unit	1 cm^2/unit
K of the soil (= 1/r)	.001 cm/hr	.01 cm/hr
-P of the soil	- 11000 cm	- 100 cm
$-\Psi_p$ of the root	- 12000 cm	- 12000 cm
Mean distance to nearest root	1 cm	10 cm

Under these conditions, the same amount of water per unit soil volume is absorbed at the two levels. How can that be?

Use Eq 6-9:

$$q_{20} = (120cm^2)(.0001cm/hr)[-12000cm - (-11000cm)]/1cm$$

$$= -12cm^3/hr \text{ absorbed per unit soil volume}$$

$$q_{250} = (1cm^2)(.01cm/hr)[-12000cm - (-100cm)]/10cm$$

$$= -12cm^3/hr \text{ absorbed per unit soil volume}$$

The example illustrates why "available soil moisture" is difficult to estimate under deep-rooting plants, and how tree roots have evolved to take optimum advantage of both the primary moisture-nutrient pool in shallow layers and the higher moisture content of deeper layers. Deep-rooting trees seldom suffer severe drought.

4. Availability of energy and water in transpiration. Leaf and root surface areas, and their ratios to the surface of the ground occupied by plants, may be taken as an index to water and energy availability. The ratio of leaf to ground area (LAI, leaf area index, L^2/L^2) varies from 1 to 10, the former in young stands and the latter in dense coniferous stands. The ratio of root surface to soil volume (L^2/L^3) decreases sharply with depth below the surface, a fact that confuses the concept of soil water availability under trees (Chap 5, Sec F). LAI's have been laboriously measured for a few forest types and estimated for stands of different type and age. LAI is used in several computer models to simulate E_t rates as a function of stand density, age, type, precipitation, etc. All considerations of the availability of energy and water in the E_t process lead hydrologists to suspect that the best indices for the consumptive use of water is the total surface area of leaves and stems exposed to the air, and the total soil moisture accessible to roots. Both are difficult to estimate.

Energy factors important in transpiration:

Radiant warming of the biosphere includes the sun and reflected or reradiated energy from clouds, soil, tree stems and air. Leaf temperature is increased only slightly above the air temperatures surrounding it, but vapor pressure gradients from leaf to air are substantially increased because e_a drops sharply with increased air temperature and convectional currents move vapor away from leaf surfaces.

Advective wind increases turbulent mixing of air within crowns of trees or plants, warming or cooling leaves, causing changes in water potentials, and stimulating closure of stomata.

Water factors important in transpiration:

Water loss from leaves causes osmotic potential of cells to become more negative, which increases the driving force ($\Delta\Psi_p/L$) from stem to leaf and stimulates closure of stomata, thus calling for more water while resisting water loss.

Water deficits in stems and roots caused by water loss from leaves do little to increase resistance to translocation but quickly apply greater "suction" at the root surface to produce more absorption from soil.

Water deficits produced in soil layers immediately around rootlets increase resistance to absorption but at the same time induce migration of water toward the root from soil deeper in the profile. When absorption is unusually rapid (as at noon on a hot day) the root-soil interface is the major impedance to absorption, translocation and transpiration. During the night, moisture will migrate toward roots, increasing $K(\Theta)$ near the root and favoring rapid absorption and rehydration of the plant before morning. During droughts, tree seedlings may be unable to rehydrate overnight but full grown trees suffer such drought chiefly on shallow soils.

5. Measurement of transpiration. The land manager is interested in the amounts of water transpired by plants on a watershed basis, and how evapotranspiration might differ under different vegetal covers. It is easy to measure transpiration from a potted plant if the pot is impervious and the plant is sealed around the stem to prevent evaporation from the soil. Such measurements have been made on many species, but they tell us virtually nothing about transpiration rates in the field. Various methods may be outlined:

Measurement on plants in pots or lysimeters.
Measurement of soil moisture depletion in plots.
Plastic tents (whole plants, even trees, enclosed in a tent; the water content of air pumped in and out is analyzed for vapor loss).
Cutting and instantly weighing branches to measure loss rate during first few minutes (usually over-estimates T).
Sap flow meters and potometers of various kinds (reveal changing rates but are not quantitative).

<u>Analysis of vapor gradients</u> and fluxes over extensive vegetal stands (difficult to verify).
<u>Analysis of the energy balance</u> of stands and plots (difficult to verify).

Measurement of transpiration on a watershed basis is an unsolved problem. Neglecting E_s and E_o, we can deduce the rates by measuring other components of the water balance and solving for transpiration:

$$T = P_g - Q - I_t - \Delta S \qquad 6\text{-}11$$

Summarizing many catchment experiments and interception studies, annual transpiration of mature deciduous hardwoods in the southern Appalachian Mountains may be roughly approximated under average annual P_g and Q. The terms in Eq 6-11 become:

T = 150cm - 70cm - 18cm - 0cm = 62cm of transpiration

About 78% of total E_t (80 cm) is transpiration under mature hardwood cover. Under mature white pine, streamflow is reduced, interception loss is increased, but annual transpiration is thought to be about the same:

T = 150cm - 52cm - 36cm - 0cm = 62cm of transpiration

About 63% of total E_t (98 cm) is transpiration under mature white pine. These estimates are good only to give the student perspective on the relative magnitude of components of annual E_t.

F. <u>Estimating evapotranspiration</u>. Land use planning often requires estimates of the main components of the water balance. Eq 2-1 clearly suggests the approaches that are available.

1. <u>Lysimeters</u> appear at first to be ideal for the purpose because each of the terms in Eq 2-1 can be measured with some accuracy on a container that has clear boundaries and no problem with leakage or surface variation. Even daily rates of E_t can be measured with precision on a weighing lysimeter, but the result cannot be extrapolated accurately. The mantle depth and surface area of a drainage basin cannot be scaled down to lysimeter-size without distorting the availability of water and energy to the evapotranspiration process. Artificial constraints on root development, soil water content and potential, and advective energy over the lysimeter have unpredictable effects on lysimeter E_t rates. Useful information on short-term variation in E_t for different crops is obtained from lysimeters, but the size of trees and the time required to grow them, defeats the use of lysimeters for estimating forest E_t.

2. Paired soil moisture plots, if the soil is fairly homogeneous and the study is carefully designed, provide comparative E_t rates for different covers, at least on a seasonal basis. Consider two adjacent plots of pine, x and c, on deep, uniformly textured soil:

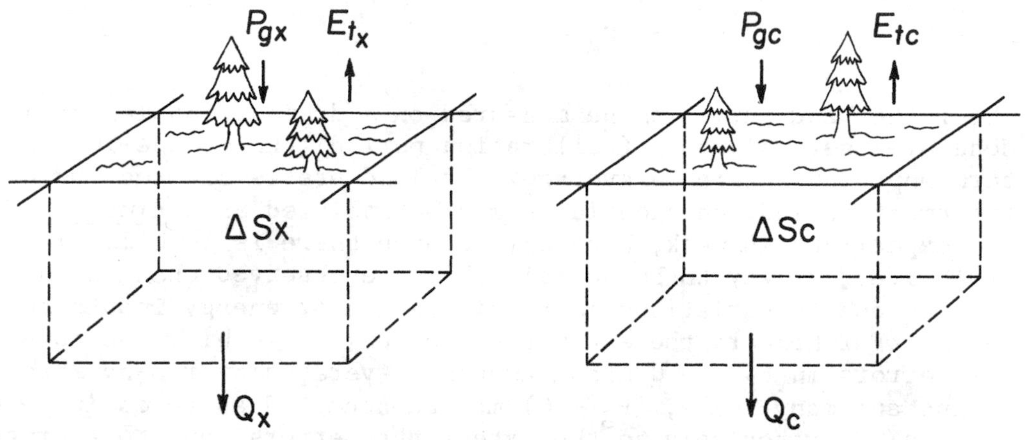

Write the water balance of each block:

$$E_{tx} = P_{gx} - Q_x - \Delta S_x$$

$$E_{tc} = P_{gc} - Q_c - \Delta S_c$$

Subtracting the two equations, we have the difference between Plot x and c:

$$E_{tx} - E_{tc} = (P_{gx} - P_{gc}) + (Q_c - Q_x) + (\Delta S_c - \Delta S_x) \qquad 6\text{-}12$$

Using the neutron scattering method (Chap 5, Sec 6), we measure the change in soil moisture storage (ΔS) under the two stands. Because the plots are close together, P_g and Q should be identical and therefore cancel out. During a calibration period before anything is done to either stand, measurements of soil moisture allows testing of this hypothesis:

$$E_{tx} - E_{tc} = \Delta S_c - \Delta S_x = \text{zero}$$

Before treatment we expect this hypothesis to be true; indeed we may assume it true if the plots are really the same.

Now we apply a treatment to Plot x; for example, the trees are sprayed with an anti-transpirant chemical to see if E_t will be reduced enough to increase soil moisture by a measurable amount. During and after the treatment period we measure soil moisture on both plots and again test the hypothesis, ($E_{tx} - E_{tc} = 0$). If it is not zero, we have an estimate of the effect of the anti-transpirant on E_t from the pine stand. The assumption that $Q_x = Q_c$ during both calibration and treatment is critical. Soil moisture should be measured to at least 3 meters under forest to gain a check on this condition, because the possibility that drainage may be changed by treatment, rainfall or irregular soil layers is a fundamental problem with this method.

3. Basin water balances seem to offer a direct approach to measuring E_t. Solving Eq 2-1 for E_t:

$$E_t = P_g - Q - \Delta S$$

The right hand terms can be measured on a drainage basin and if done over several years (calibration period) the change-in-storage term approaches zero on average. If land use is altered during a treatment period, changes in E_t may be analyzed as a time trend. The experiment is weak, however, because there is no climatic control, i.e., no way to be certain that the observed changes in E_t are not due to variations in precipitation or energy inputs to the basin. Furthermore the estimates are subject to bias due to systematic errors in P_g and Q measurements. Averaged over many river basins and many years, $(P_g - Q)$ may be accepted as an estimate of regional E_t rates only to the extent that errors tend to average out. The single drainage basin is a poor experimental tool for determining the effects of land use and vegetal cover on E_t.

4. Paired catchment method. To overcome the experimental difficulties with basin water balances, forest hydrologists devised the paired watershed experiment. The first was the classical Wagon Wheel Gap Experiment in Colorado, jointly carried out by the U.S. Forest Service and Weather Bureau between 1909 and 1919 (Bates and Henry 1928). The purpose of such experiments is not so much to determine the absolute values of E_t, but the absolute differences in E_t among two or more forest cover types. Fig 6-4 shows the method, simplified for clarity. Two nearby or adjacent catchments (25 to 50 ha is the best size) are selected for gaging on the basis of their similarity in size, cover type, aspect and suitability for streamflow gaging. Using subscripts x and c as in the paired soil plot method, the experiment may be diagrammed:

	BASIN x	BASIN c
CALIBRATION PERIOD	NO CHANGE	NO CHANGE
TREATMENT PERIOD	TREATED	NO CHANGE

With the paired catchment design, we need make fewer assumptions in Eq 6-12 than we did with soil plots. P_g and Q are measured on both basins; basin storage is assumed to reach the same level during spring recharge or fall low flow, and thus the term $(\Delta S_c - \Delta S_x)$ tends toward zero on an annual basis.

The success of the method is based on the high correlation that normally exists between Q_x and Q_c (the annual discharges from the

CALIBRATION PERIOD

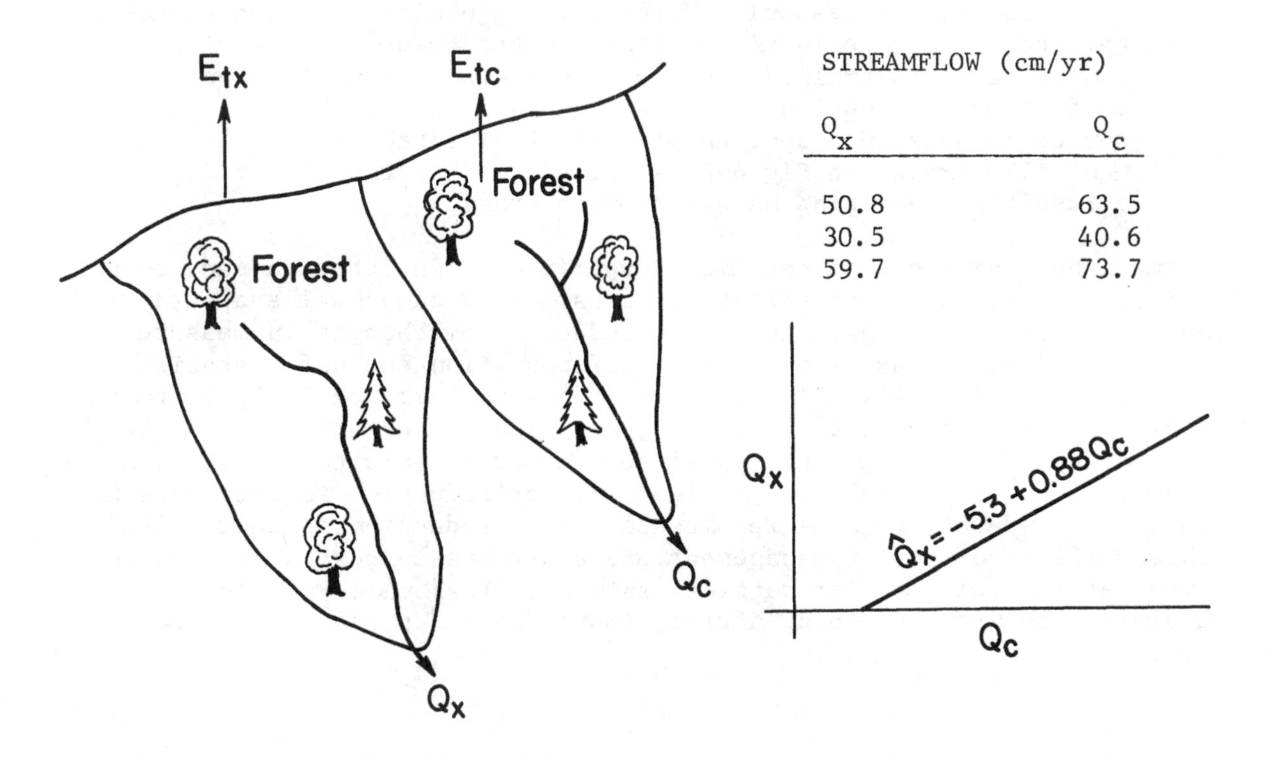

STREAMFLOW (cm/yr)

Q_x	Q_c
50.8	63.5
30.5	40.6
59.7	73.7

TREATMENT PERIOD

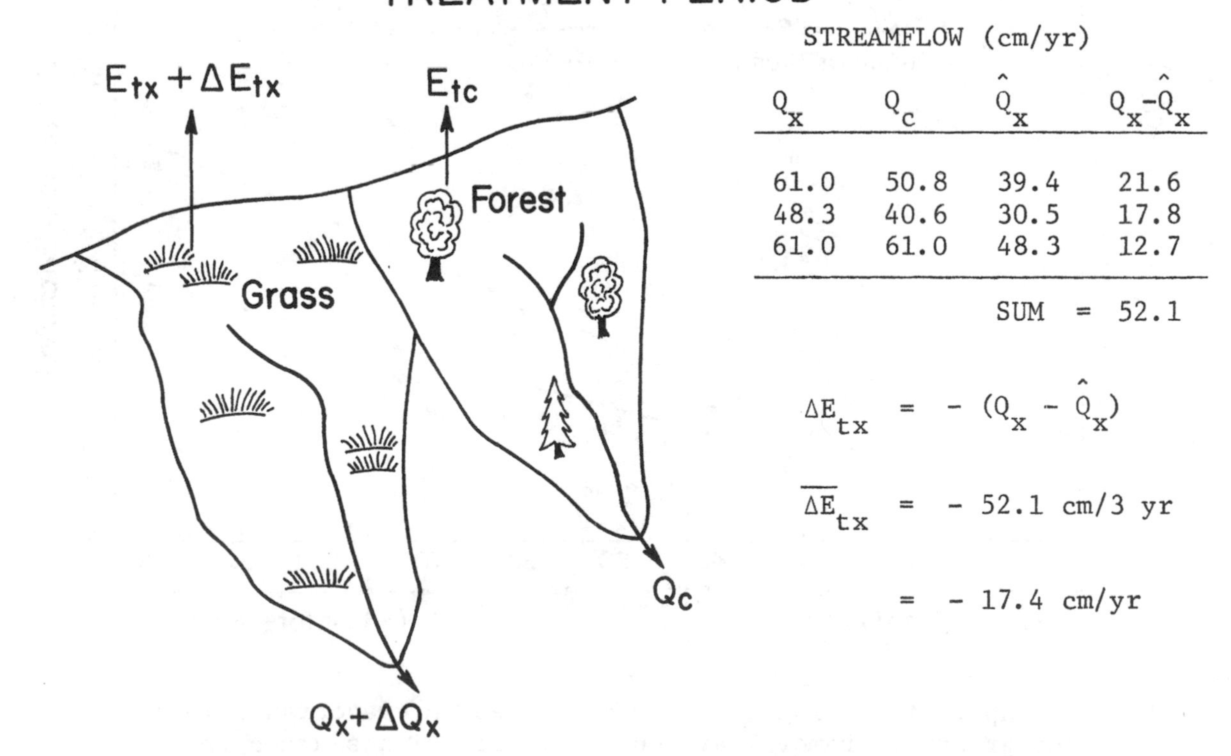

STREAMFLOW (cm/yr)

Q_x	Q_c	$\hat{Q}_x$	$Q_x-\hat{Q}_x$
61.0	50.8	39.4	21.6
48.3	40.6	30.5	17.8
61.0	61.0	48.3	12.7
		SUM =	52.1

$$\Delta E_{tx} = -(Q_x - \hat{Q}_x)$$

$$\overline{\Delta E}_{tx} = -52.1 \text{ cm/3 yr}$$

$$= -17.4 \text{ cm/yr}$$

Fig 6-4. A simple example of a paired catchment experiment to determine the effect of converting forest to grass on evapotranspiration and water yield. The analysis does not assume that basin leaks are zero, but assumes that if leaks occur, they are constant with respect to differences.

two watersheds) when the vegetal cover is the same. This correlation is evaluated during the calibration period by regression analysis and the resulting "prediction equation" is used to determine the change in yield after the treatment. Since climatic changes are "controlled" by comparison with the undisturbed catchment c, we may conclude that changes in Q_x after the treatment were due to changes in evapotranspiration from catchment x. The experiment illustrated in Fig 6-4 is solved to show the change in E_{tx} as a result of replacing hardwoods with grass.

Summarized from about 75 catchment experiments conducted in many countries, Fig 6-5 shows the effect of forest cover on annual evapotranspiration rates. Changes in E_t are deduced from changes in measured water yield of basins after various afforestation and deforestation treatments. Virtually all experiments show an increase in E_t as forests grow and a decrease in E_t as forests are clearcut or thinned. A 10% reduction in conifer or eucalypt forest cover appears to result in about 4 cm reduction in annual E_t, while a similar reduction in deciduous hardwoods in temperate regions results in 2.5 cm reduction in annual E_t. These effects of forest management are a serious concern in regions of high water requirement or marginal rainfall (less than 80 cm/yr). Examples are Great Britain, Africa, Australia and parts of southwestern U.S.

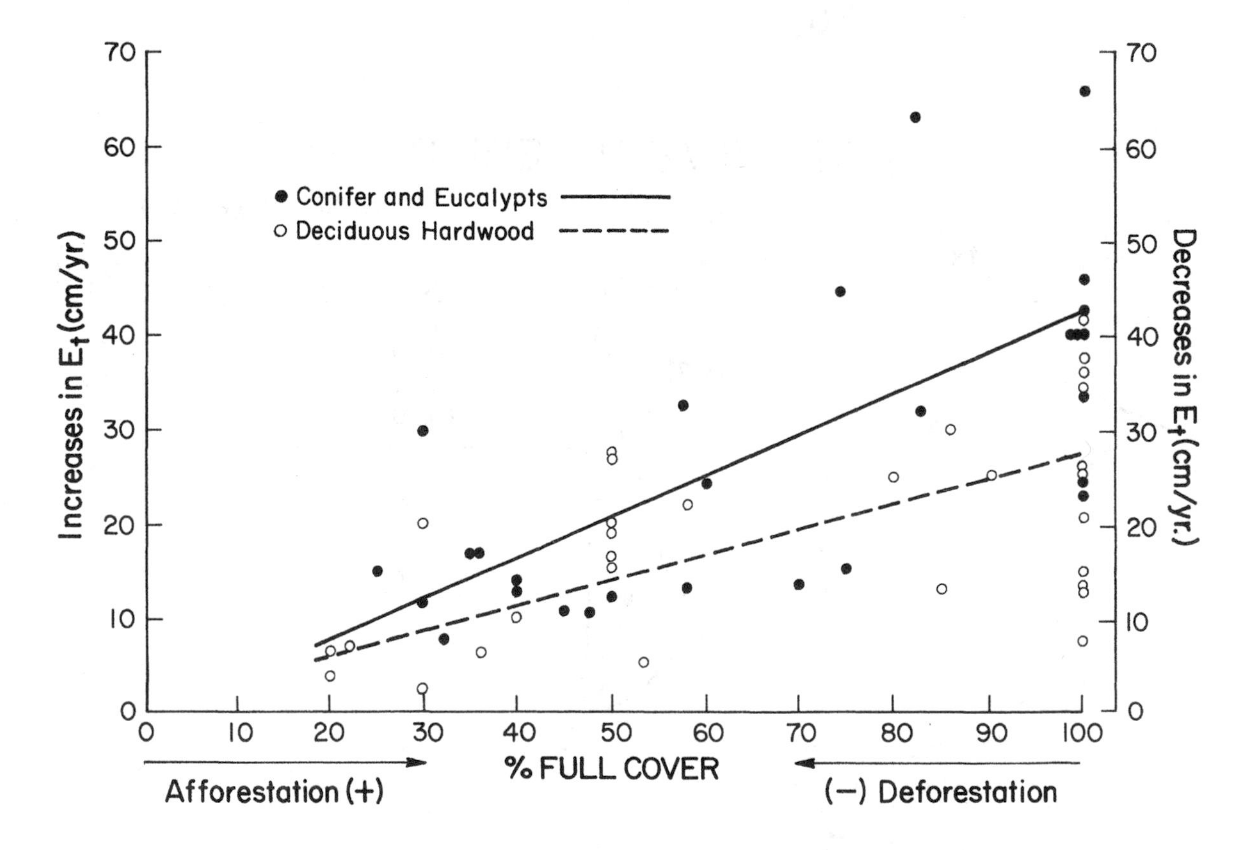

Fig 6-5. Approximate increases or decreases in evapotranspiration due to forest growth or removal are shown by trend lines drawn through the results from 75 experiments around the world (from Bosch and Hewlett 1981).

5. Meteorological approaches (energy and vapor balances). Much research has gone into the development of theoretical equations for estimating E_t from solar radiation, air and water temperatures, vapor pressures, wind speed, precipitation and other climatic variables. The most widely used has been Thornthwaite's empirical formula based on mean monthly temperatures and day length. He begins by calculating PE_t in mm/dy for given monthly temperatures:

$$PE_t = 0.53b\ (10\ t_a/I)^a \qquad 6\text{-}13$$

where I is a weighted mean annual temperature, t_a is mean monthly air temperature, a is a cubic function of I, and b is a correction for day length. The coefficients were estimated by regressing these variables on precipitation and streamflow records from plots and small basins. The formula is needlessly complex; much simpler equations give similar answers. One is that of Holdridge (1962) which uses only daily mean temperature:

$$PE_t = 0.161\ b\ T_a \qquad 6\text{-}14$$

PE_t is in mm/dy, b is a correction for day length and T_a is daily mean temperature in °C (maximum plus minimum temperature divided by 2). Only zero or positive values of °C are used; set minus values equal to zero in computations. The correction for day length is a function of season and latitude. This simple method will serve to distribute annual values for $(P_g - Q)$ measured on actual basins proportionately through the year. It will, however, tend to allocate daily E_t rates smaller than actual ones during the winter months, and larger than actual during summer.

Formerly soil water budgets and monthly streamflow were predicted from such methods but they proved highly unreliable. To quote Lee (1978), "There is no adequate method of predicting E_t in the biosphere based on simple weather-element data (because) E_t is a phenomenon with atmospheric, surface, subsurface and biological control mechanisms." At present the safest approach to its estimation for the purpose of making decisions about forest and wildlands is undoubtedly the watershed water balance (Eq 2-1) and the catchment experiment. Whenever regional controversy arises from conflicting claims about the consumptive use of different crop and forest practices, there is no substitute for a locally-conducted catchment experiment to settle the issue.

G. Control of evapotranspiration is sought to increase streamflow (downstream water supply) or to conserve soil moisture for further plant growth (on-site water supply). The heavy water requirements of forest, particularly conifers and eucalypts, are already limiting the land available for forestry in some regions. In Great Britain, some cities have passed restrictions on afforestation of municipal watersheds. In South Africa laws have been passed to prevent afforestation of certain mountain catchments and water courses. Efforts to control evapotranspiration are routine in dry-land agriculture; much of the eastern Montana and Dakota wheatlands is a pattern of wide strips, one left fallow and one planted, because one year's precipitation is normally not enough to grow a crop of wheat. A number of ways are in use or under consideration for reducing E_t or increasing productivity on site.

Fallow the land for one or more seasons to carry over some soil moisture for the next growing season.

Cut, mow or otherwise eliminate plants, leaving soil bare or covered with vegetal debris.

Defoliate plants with chemicals which cause the abscission layer to form in the petioles of the leaves.

Substitute species or forest types which have a lower total $(T + I_t)$, and thus favor moisture conservation.

Spray leaves with anti-transpirants to cause temporary closure of stomata. (Since transpiration is a leakage process, that is, an unavoidable consequence of the provision of stomata for CO_2 exchange, reduction of T by stomatal closure should conserve water and do little harm.)

Shade the plants, as is done in shade-grown tobacco and tea.

In semi-arid regions, it may be possible to reduce E_t by elimination of forest or brush cover but nevertheless provide no appreciable water for other uses because the conserved water merely evaporates from the bare soil where the plants stood. For example, experiments have shown that eliminating pinon and juniper trees in Arizona locally reduces E_t but provides no detectable increases in streamflow or groundwater supplies. However, such treatments may shift on-site consumptive use to more productive grasses or forbs. The economics of E_t conservation is well developed in irrigation practice, but not yet in forest, range and wildland management.

Further readings.

Bosch, J. M. and J. D. Hewlett. A review of catchment experiments to determine the effect of vegetation changes on water yield and evapotranspiration. J. of Hydrol., Vol. 55, 1982 (in press).

Helvey, J. D. A summary of rainfall interception by certain conifers of North America. IN: Biological Effects in the Hydrological Cycle, Proc. 3rd Int. Semin. Hydrol. Professors, Purdue Univ., Indiana, pp. 103-113, 1971.

Kramer, P. J. Plant and Soil-Water Relationships: A Modern Synthesis. McGraw-Hill Book Co., N.Y., 482 pp., 1969.

Lee, R. Forest Microclimatology. Columbia Press, N.Y., 276 pp, 1978.

Monteith, J. L. Principles of Environmental Physics. American Elsevier Publish. Co., N.Y., 1973.

Penman, H. L. Vegetation and Hydrology. Commonwealth Agricultural Bureau, Farnham Royal, England, 1963.

Slatyer, R. O. Plant-Water Relationships. Academic Press, N.Y., 366 pp, 1967.

Sopper, W. E. and H. W. Lull, (Ed.). International Symposium on Forest Hydrology: Session IV, Forests and evapotranspiration. Pergamon Press, Oxford, pp. 373-494, 1967.

Problems.

1. A 1000-acre mature hardwood forest watershed is the source area for a municipal water supply. Compute the inches and acre feet of water evaporated as total interception loss, using the calendar year's rainfall records provided. Convert to gallons and estimate how many day's supply this would be for a city of 50,000 population with an average daily per capita use of 180 gallons.

2. The total above-ground mass of vegetal material on a certain forested watershed is 2 x 10^5 kilograms per hectare. The specific heat of this material is about 700 calories per kilogram/°C. A light rain cools the vegetation from 30° to 20°C. Assuming that all this heat energy was used to evaporate intercepted water, compute the millimeters of water evaporated.

3. Compute (in meters) the total water potential Ψ and the plant water potential Ψ_p in the top of a non-transpiring redwood tree 80 m high when the tree is standing in still, saturated air, as during early morning hours under foggy conditions. Why doesn't the water drain out of the tree?

4. The following table contains monthly rainfall and streamflow data from an experiment on two forested catchments x and c in the southeastern Piedmont. The forest on catchment x was clearcut during the winter of 1974-75. Analyze the data for monthly changes in summer water yields. What important assumptions have been made in drawing your conclusions?

MONTH	CALIBRATION (1974)			TREATMENT (1975)		
	Monthly P_g	x	c	Monthly P_g	x	c
		----------	cm	---------		
May	8.1	0.6	0.7	16.1	7.6	3.5
June	4.8	0.3	0.4	8.8	3.0	0.8
July	15.0	1.8	1.9	10.8	1.9	0.5
August	20.5	3.6	3.3	12.2	6.3	2.2
September	6.2	0.5	0.5	8.7	2.4	0.5

5. One year's record of monthly rainfall and streamflow in mm will be handed out in class. Plot the <u>apparent</u> monthly evapotranspiration (P_g - Q, by month). Determine mean monthly temperature and the day length correction factor from local meteorological references. Use Eq 6-14 to compute monthly potential evapotranspiration in mm; plot on the same graph. Discuss the errors between monthly and annual PE_t and apparent E_t. Day length factor b = (daylight hours)/12.

CHAPTER 7

SURFACE WATER, THE HYDROGRAPH AND THE RUNOFF PROCESS

A. Surface waters. Only a tiny fraction of the world's fresh water is found on the surface of the land at any one time. These rivers, lakes, reservoirs, ponds and swamps constitute most of our usable supply of water, although the exploitation of ground water has increased greatly since the development of efficient pumps. Many problems relating to the quantity, quality, timing and energy disposition of water develop while water is on the land; consequently a large part of the subject of hydrology relates to the processes involved in the storage and movement of water over the surface of the earth. Many specialized fields exist:

Hydraulic engineering
Flood prediction, protection and control
Water pollution control and sanitary engineering
Inland water transportation
Inland and anadromous fisheries management
Reservoir and hydroelectric power management
Land and water conservation (soil erosion)
Water recreation management
Water fowl and other wildlife management
Land drainage and irrigation
Avalanche control and snowfield recreation
Water supply engineering (includes ground water)
Limnology and glaciology

Measurement of water on the land surface consists chiefly in (1) the survey of the depth, area and other features of rivers, reservoirs and ice fields, and (2) the gaging of rates of flow in rivers, reservoirs and small channels. We have dealt with snow pack measurement in Chap 4 because of its close association with precipitation measurement. Streamflow gaging is an activity often associated with hydrology in the mind of the layman.

1. Water stage. The simplest and locally most important measure of water flow in channels is the stage, or height of a water surface above a datum, usually the bottom or the bank of a river or lake. Flood stage refers to a high water level that does damage to human property or which overflows the normal banks of a stream. The maximum flood stage along a river usually occurs at the time of maximum discharge following a large storm. Staff gages are used to indicate the stage of a stream; these are painted rods or

boards scaled so that they can be read from the bank or from a bridge.

An improvement, to record maximum flood stage when no one is present, is the crest gage. There are various kinds but all leave a mark at the high water stage until an observer can record it. Only the largest peak in the interim leaves a mark.

Valuable estimates of the height of a recent flood flow can be deduced from water or mud marks on trees or bridge piers, but caution is required in their interpretation. Such marks are most reliable when they are found in an eddy or back-water where upsurges in a swift current are not likely to leave marks higher than the peak stage. Debris caught in the branches of small trees or brush may be an unreliable high water mark because the branches and the debris will rise as the force of the current relaxes against the stem. Recorded notes or a mark and a date cut on a large tree may guide future planning at the locality, particularly if the observer has good reason to believe that he has just witnessed an exceptional event.

Many kinds of clock-driven instruments that record stage for weekly or monthly periods have been developed (water level recorders). The U.S. Geological Survey (USGS) alone operates some 13,000 of these in the United States on streams draining areas up to tens of thousands of square kilometers. Some records go back 100 years, but there is still not enough of this information to satisfy all users. The recorders are expensive and require frequent servicing. The older ones produced an analog trace of the water stage on rolls of paper; recent ones (called analog-to-digital recorders) punch stage values into long rolls of paper tape which are later automatically transferred to cards or magnetic tapes. The latter allow complete electronic processing of records. Some commonly used water level recorders are the FW-1 Friez pen recorder (daily or weekly charts), the A-35 Stevens pen recorder (six-month chart) and the ADR Fisher-Porter punch-tape recorder (six-month tape).

Another type of water level recorder is called a "bubbler gage." Its advantage lies chiefly in the fact that the recording device and housing can be located some distance up the bank, well out of the path of flood waters. Through a buried tube, a small orifice anchored at the zero datum in the bottom of the stream is connected to a tank of nitrogen gas in the gaging house. The gas is allowed to trickle very slowly into the water. The back pressure upon the slowly escaping gas is equivalent to the pressure head (P) due to the depth of the water over the orifice. The pressure differences in the tube when the orifice is exposed to the atmosphere and when it is emersed in the water is recorded as an indirect,but quite accurate, measure of the head of water over the bottom of the stream.

The stage, usually called head (h) in hydrologic work, measured by the water level recorder is only part of the information needed to determine stream discharge (Q) in volume per unit time (L^3T^{-1}).

2. The hydrograph. When head (h) or discharge (Q) is plotted over time in minutes, hours or days, the resulting curve is a hydrograph. Total flow, seasonal distribution of flow, daily flow, peak flow, minimum flow and the frequency of various critical flow rates are all computed from the hydrograph. Figure 7-1 diagrams a hydrograph produced by a small basin following a four-hour storm; however, few actual hydrographs are so regular in form.

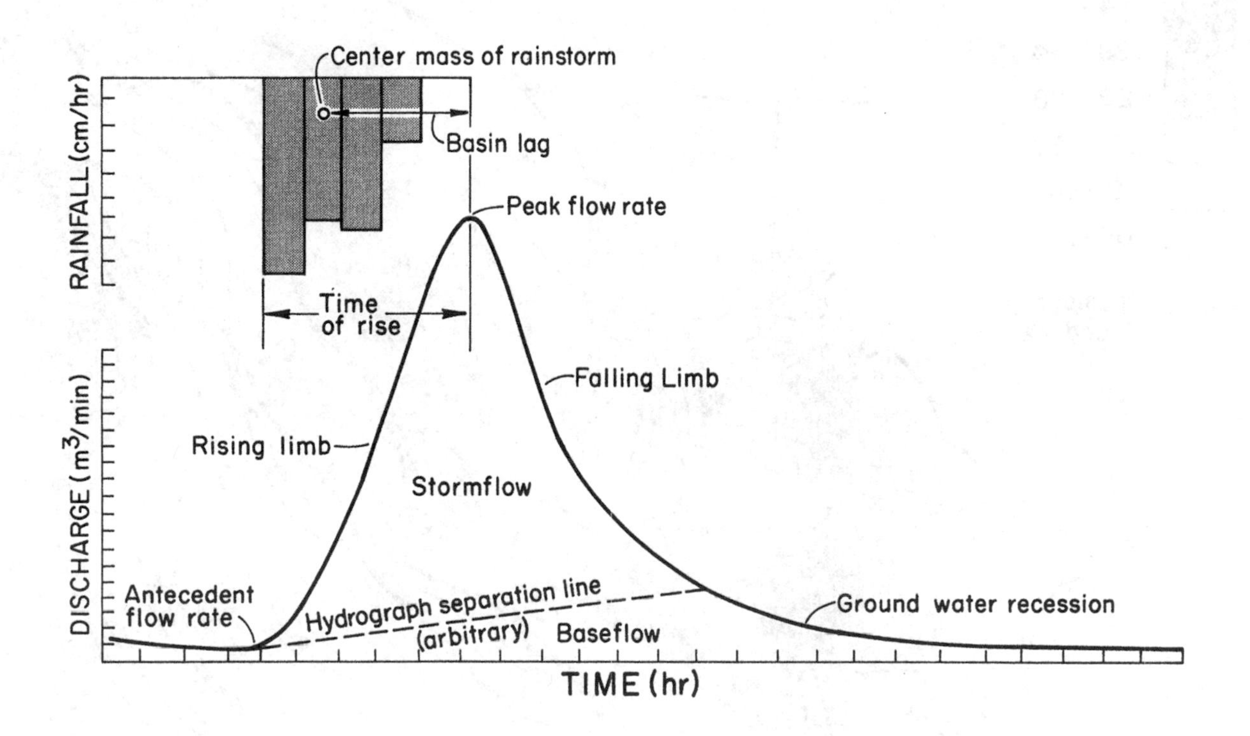

Fig 7-1. This diagram defines terms normally used to describe a storm hydrograph.

3. Components of flow vary with size of rainstorms and with antecedent basin wetness. The separation of stormflow from baseflow depends entirely on the judgement of the hydrologist; such flow classifications must be arbitrary because the source of water is not revealed in its hydrograph.

The hydrologic response of land refers to the "flashiness" of stormflow following rainstorms. A simple way to express response is to classify stormflow (Q_s) quantitatively and divide by the storm rainfall (P_g) that produced it:

$$\text{Hydrologic Response} = Q_s / P_g$$

$$\text{Mean Response R} = (\overline{Q_s / P_g})$$

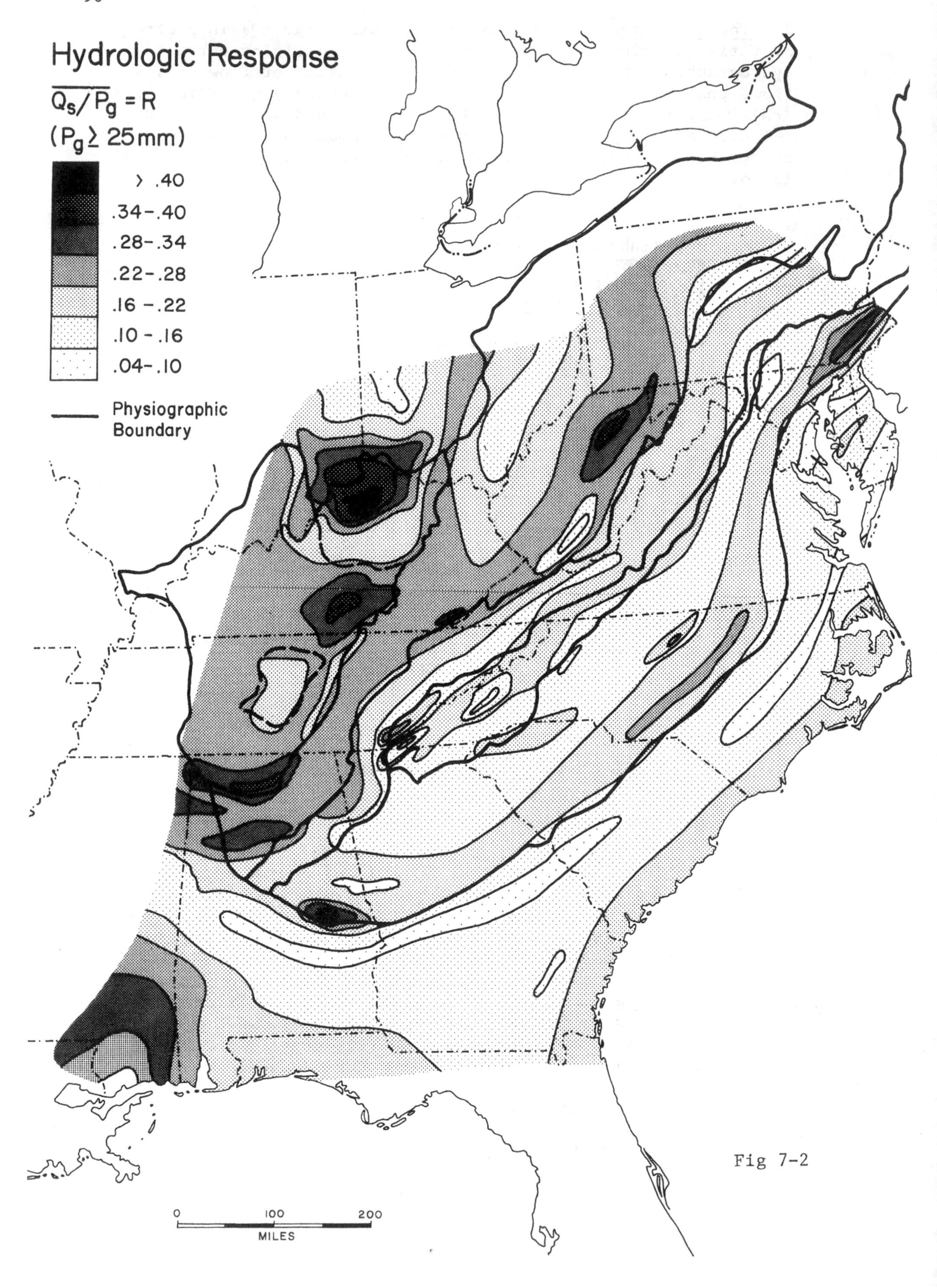
Hydrologic Response
$\overline{Q_s/P_g} = R$
$(P_g \geq 25 mm)$
> .40
.34-.40
.28-.34
.22-.28
.16 -.22
.10 - .16
.04-.10
Physiographic Boundary
0
100
200
MILES

Fig 7-2

Rainstorms smaller than 2.54 cm (1 inch) produce small stormflows that do little damage, so it is convenient to compute mean response R only on storms greater than 2.54 cm. There are about 10 to 20 such rains per year in eastern U.S. Computed from 3000 water years of hydrologic record, Fig 7-2 shows that mean response R of headwater basins varies by physiographic province. The weighted average R over eastern U.S. is about 0.20, that is, about 20% of a typical rainfall becomes stormflow. A typical rain is about 5 cm in one or two days. Over twice as much stormflow (direct runoff) is produced in the Cumberland Plateau as in the Upper Coastal Plain. The Plateau is composed mostly of sandstone caps over limestone, and shallow soils on steep slopes. The Louisville Basin in the bend of the Ohio River exhibits the flashiest response in the East. Because these soils are shallow, basin storage is low. The Upper Coastal Plain, on the other hand, is composed of deep, permeable, sandy soils. A strip of the Coastal Plain along the fall line of the Piedmont, known as the Sand Hills, delivers only 4 percent of a typical storm as stormflow. The Lower Coastal Plain has high water tables which reduce storage capacity. The Piedmont is intermediate; its soils are generally deep with good surface infiltration but have impeded percolation in the B horizon. Although the mountains are steep, the coarse granitic soil, forest cover and deep mantles prevent excessive response except above 1200 m elevation.

It is clear from the map that mean hydrologic response is controlled more by geology than by land use. Because of early overemphasis of infiltration as a limiting factor in the runoff process, land use has tended to receive more blame than it deserves in the production of stormflows and floods.

Diurnal fluctuations in streamflow occur during summer as a result of daily evapotranspiration along or near the stream channel. Clearing vegetation along streams reduces these fluctuations. The diagram at the top of the next page shows one day's hydrograph from a small temperate zone watershed. The expected yield during the 24 hours from 10 am to 10 am is about 15 percent greater than the actual yield. The reduction is ascribed to evapotranspiration by riparian vegetation.

4. Discharge measurement. Not until the 18th Century did Bernoulli demonstrate that the volumetric flow of water in channels (Q) was the product of velocity (V) and the cross-sectional area (A) through which the water flowed:

$$Q = AV \qquad \text{UNITS: } L^3T^{-1} \qquad 7\text{-}1$$

Eq 7-1 is deceptively simple because it is not easy to determine the average velocity of water in a stream. The width, depth, slope and frictional drag of the channel are all involved as the potential energy of the water is dissipated flowing from a higher to a lower level in the earth's gravitational field. The frictional drag of the bottom and sides of the channel reduces water velocity

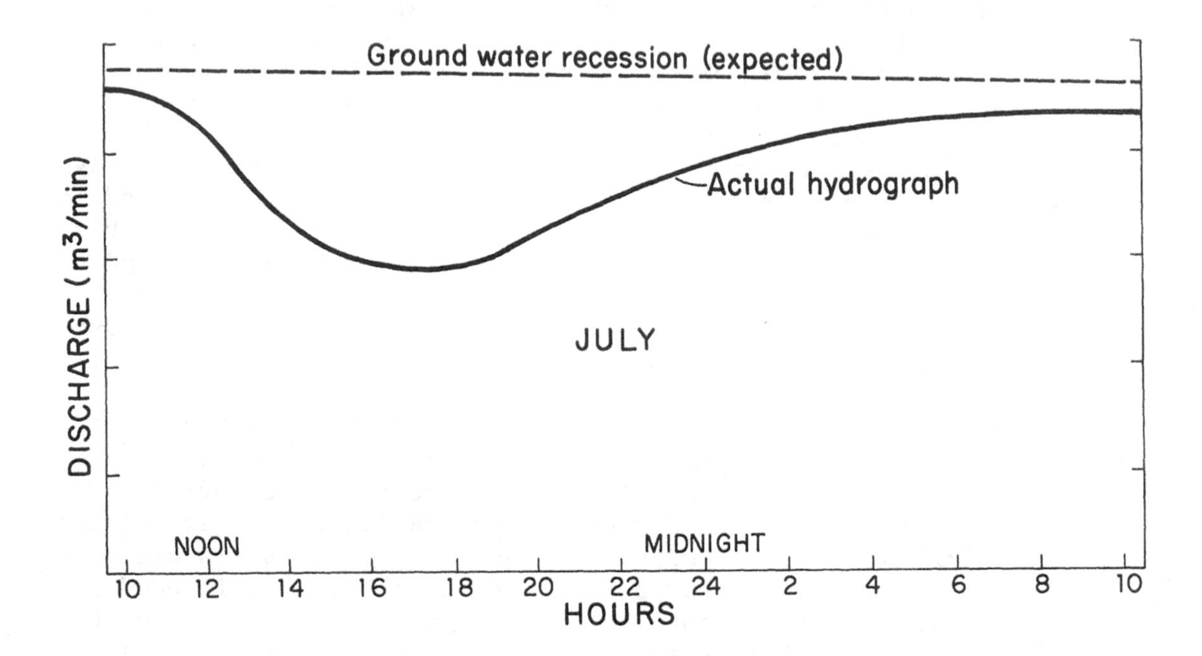

from a maximum in the middle of the stream to zero at the exact points of contact between the water and the channel bottom. Measured across the stream at right angle to the direction of flow, the line that connects these points is called the wetted perimeter (W_p). The ratio

$$r = \frac{A}{W_p} \qquad 7\text{-}2$$

is known as the hydraulic radius. The hydraulic radius is one of two main factors (the other is channel slope) determining the capacity of an open channel to carry water at different stages. For a given cross-sectional area A, a large hydraulic radius corresponds to a small wetted perimeter and a relatively high average water velocity in the channel. Thus channels with very irregular bottoms, as in the mountains, may often require cross-sectional areas larger than smooth U-shaped channels, as in the coastal plains, to pass the same discharge Q. For this reason, the mean velocity of a splashing mountain stream is frequently slower than the velocity of a coastal plain stream having the same cross-sectional area but a larger hydraulic radius. We see that slope is not the only factor affecting stream velocity. The mountain channel is steeper, but the potential energy of its water is partly dissipated in turbulence and in work performed against the channel banks. Furthermore, the riffles are interrupted by pools and eddies which also serve to drop the average velocity. In the coastal plain channel with its flatter gradient, some of the potential energy is allowed by the smoother channel to dissipate in accelerating the water to a higher velocity. Neglecting free water falls, the fastest water velocities, up to 6 m/sec, are measured in the middle of large rivers near the sea. Average velocity in a small mountain stream is less than 0.5 m/sec.

Gaging methods. For this elementary discussion of flow in streams, we shall assume what the hydraulic engineer calls steady, uniform flow. *Steady flow* in an open channel means that the discharge Q is not changing with the passage of time; during the hour or so required to measure Q in a stream we may assume this condition. At the same time, if we select a reach of the stream that is fairly straight and uniform in gradient over a distance of 20 to 30 m we may assume that the flow is *uniform*, that is, the cross-sectional area through which the water is flowing is not changing appreciably above and below the stream cross-section we wish to use for measurement.

Fig 7-3 shows how Q is related to stage (h), but only as long as the channel bottom and banks remain stable. The product AV increases as the stream rises, and decreases as it falls in such a way that for every value of h there is only one value of Q. (There are exceptions: if the stream is partially dammed by constriction or obstruction below the gaging section, a rising stage of a certain value may represent a somewhat lower discharge Q than a falling stage of exactly the same value. This is one of the reasons for careful selection of a uniform reach.)

Velocity is usually measured by current meters, the most common type being the Price current meter. The Pygmy current meter is a modification for use in small streams. A rotating cup-propeller is emersed in the stream; each rotation of the propeller is counted and velocity is computed from a calibration equation furnished with the meter (Fig 7-4).

In small streams, one man in waders, with current meter, stop watch and notebook in hand, anchors a measuring tape to a stake on the bank and records distances, depths and velocities as he wades across the stream. The stream is normally divided into about 10 subsections of differing widths as illustrated in Fig 7-3. Velocity is measured within each subsection in accordance with either the ".6th depth" or the ".2th plus .8th depths" rule. If the water in the subsection is deeper than 0.3 m, the latter rule is used; less than that, the former. If the hydrologist is concerned about the triangular shape of the two end sections, he does not resort to detailed measurement of the triangles, but rather increases the number of subsections to attain an acceptable level of accuracy more efficiently.

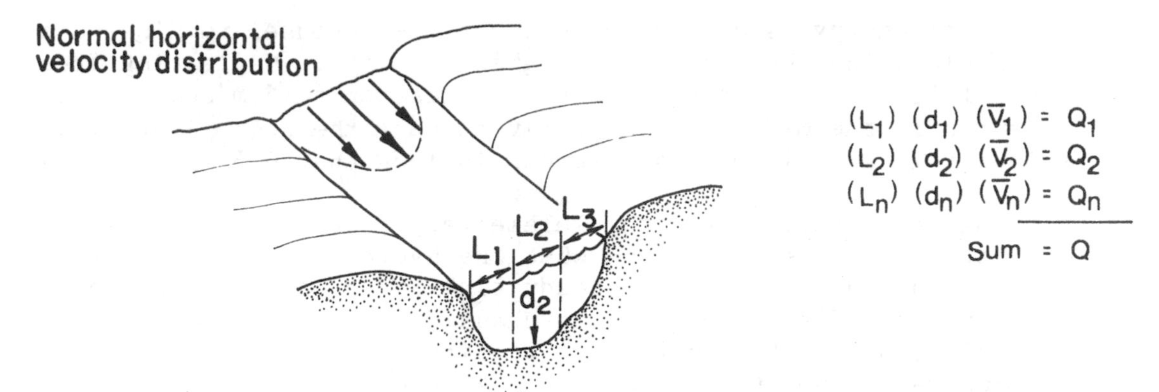

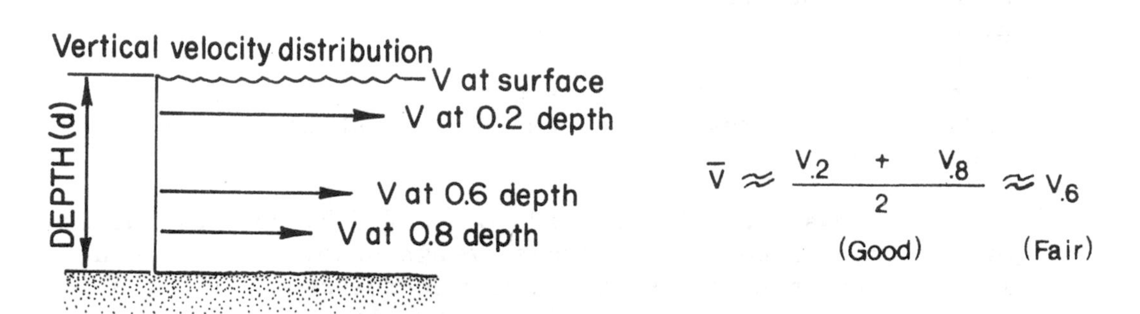

Fig 7-3. Diagram showing channel flow and how to measure it.

Fig 7-4. Diagram showing a Pygmy current meter in operation. The observer records the clicks per minute indicating the velocity of the water.

Example: Determine stream discharge Q using this data:

Dis- tance	Depth	Measured at	Velocity	Cross- section	Flow
m	m	tenths	m/sec	m^2	m^3/sec
2.0	0	-	-	-	-
2.4	0.5	6	0.3	0.6	0.18
4.0	1.0	2 & 8	(1.0 + .2)/2	1.8	1.17
6.0	1.4	2 & 8	(1.3 + .4)/2	2.1	1.78
7.0	0.4	6	0.2	0.5	0.10
7.7	0	-	-	-	-
				Q =	3.23 m^3/sec

An individual measurement of Q by this method is classified as "good" if 95% of the results lie within 5 to 10 percent of the true discharge, and "excellent" if within less than 5 percent. Measurement of stream discharge on small streams is seldom better than plus or minus 10 percent of the true value.

A simple but useful method for estimating discharge in small streams requires only a watch and a tape. Velocity at the surface of the water in the middle of a straight reach of stream can be estimated by timing the downstream movement of a chip or twig. Since this is not the average but close to the maximum velocity in the stream, it must be multiplied by a rule-of-thumb coefficient (about 0.75) to estimate $\overline{V}$. Cross-sectional area may be quickly measured with the tape to estimate discharge. Accuracy will be "poor" but the true discharge should lie within 20-25% of the estimate.

Another simple method gives precise estimates but is useful only in small streams or springs which can be diverted from a rock ledge or small dam into a calibrated bucket. Liters of water caught in a measured number of seconds is easily converted to discharge rate in any units.

Discharge integration. If h is continually recorded on a clock-operated chart, the result is a trace representing the change in stage over time (Fig 7-5). Q is measured at several values of h to produce a rating curve which is used to generate a hydrograph of discharge (Fig 7-5). A common job of the field hydrologist is to prepare and continually update the rating curve for each stream gaging station. Scouring and deposition is such a problem in rivers the size of the Mississippi that rating curves are unreliable, so depth and velocity must be measured frequently.

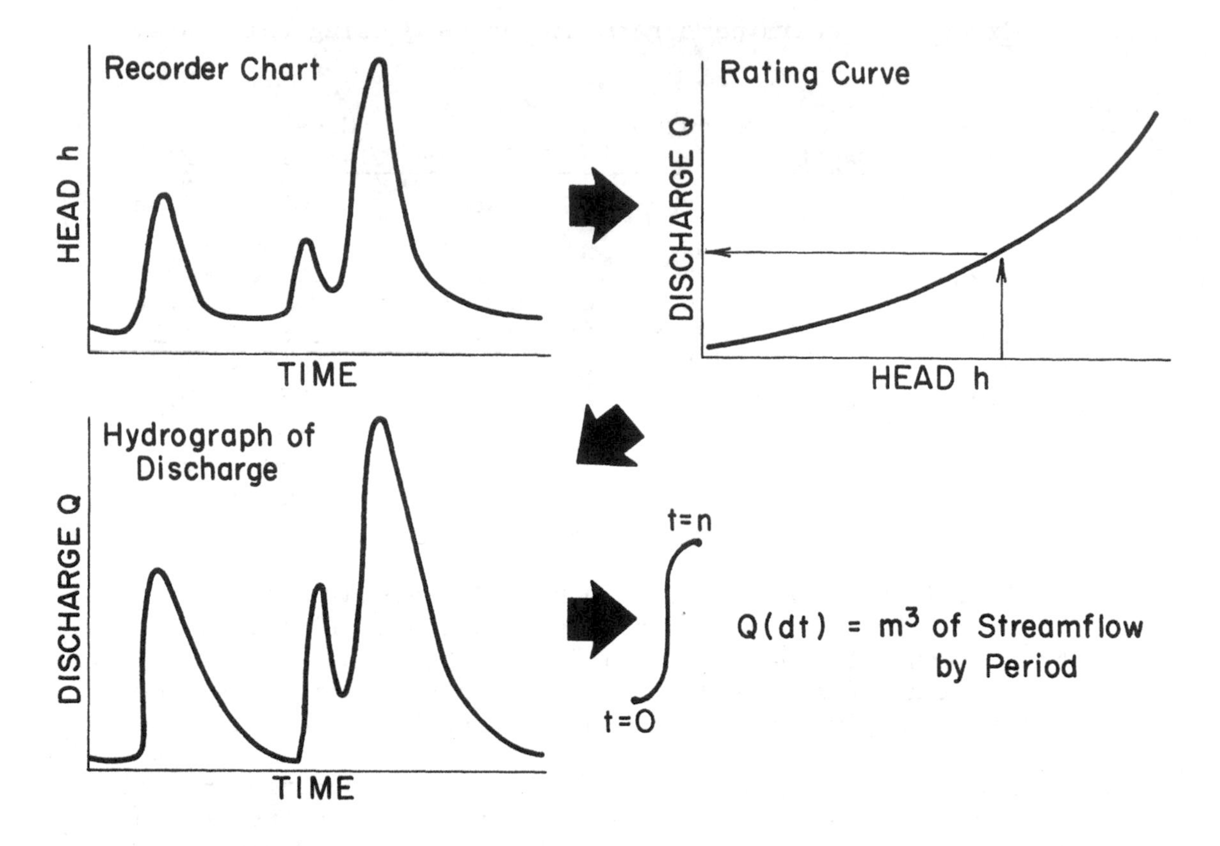

Fig 7-5. The computation of streamflow amount by storm periods or by time periods from water level records.

5. Discharge estimation by formula. A widely used method for estimating discharge, particularly flood peak discharge from highwater marks, is the Manning-Chezy formula, named for Chezy who first proposed it and Manning who modified it for easier use about 1890:

$$Q = \frac{1}{n} A r^{2/3} s^{1/2} \qquad 7\text{-}3$$

Q is stream discharge in m^3/sec; A is cross-sectional area in m^2; r is hydraulic radius in m; s is slope gradient of the water surface (dimensionless); and finally "Manning's n" is a channel roughness factor, in units $TL^{-1/3}$. Manning's n was found by experiment to vary from 0.02 in smooth channels to 0.15 in rough weedy channels. (Substitute 1.49/n for 1/n if ft and sec are used.)

Slope, area and hydraulic radius can be estimated along a straight reach of channel using high water marks, or the actual surface of the flowing water, and previously tabulated stage and area relationships. Tables of n can be found in nearly all hydraulic manuals. The formula reveals the inverse relation between discharge and channel roughness (n) and the direct relation of discharge to hydraulic radius (explained before) and to channel slope (actually water surface slope).

> Example: From high water marks it was determined that a flood through a short reach of channel where we wish to locate a culvert had a water surface slope of 0.1 m in 10 m. At that flood stage, the cross-sectional area (A) was 1.5 m^2, and the wetted perimeter W_p was 2 m. The channel was winding above and below the reach with some pools, riffles and a stony bottom, but very little vegetation on the banks or bottom. How many m^3/sec must the culvert carry to handle floods up to that size? From a handbook of hydraulics or hydrology we find that the value of n should be 0.05.

$$\text{Hydraulic radius } r = A/W_p = 1.5m^2/2m = 0.75m$$

$$\text{Discharge } Q = \frac{1}{n}(1.5m^2)(.75m)^{2/3}(.1m/10m)^{1/2}$$

$$= \frac{1}{.05}(1.5)(.825)(0.1) = 2.48m^3/sec$$

The Manning-Chezy formula is by no means exact; no equation has yet been derived from hydraulic theory that will account for all variations in slope, sediment loads, frictional drag, head losses and turbulence that control the flow of water in natural channels. However, use of the formula in planning and elementary design has proved more satisfactory than estimates based on visual survey of the site.

6. Gaging stations. A typical USGS gaging station makes use of a corrugated stand-pipe installed to a level below the river bottom and connected to the river flow by a small underground pipe. The stage in the river is the same as the elevation of the water in the corrugated stand-pipe (stilling well). A water level recorder is mounted on a shelf above expected high water. The river banks and channel are undisturbed, giving the name uncontrolled section to this type of station. Selection of the reach is again important to insure a uniform approach and departure from the section.

Various problems arising out of discharge measurement and the need for highly accurate discharge records for hydrologic experiments led to the development of controlled sections (weirs and flumes). A weir is a device for controlling the flow of water for various purposes, one being to measure the discharge Q. A flume is a similar device, modified to keep the water moving for some distance in a measurable cross-sectional area A. Controlled sections were devised to minimize errors in determining the relation of Q to h. Fig 7-6 shows one type, a 90-degree, V-notch, sharp-crested weir. In this case, the blade of the weir is designed to relate Q to h by an equation derived from hydraulic theory, verified in hydraulic laboratories. From the geometry of triangles, we see that A is:

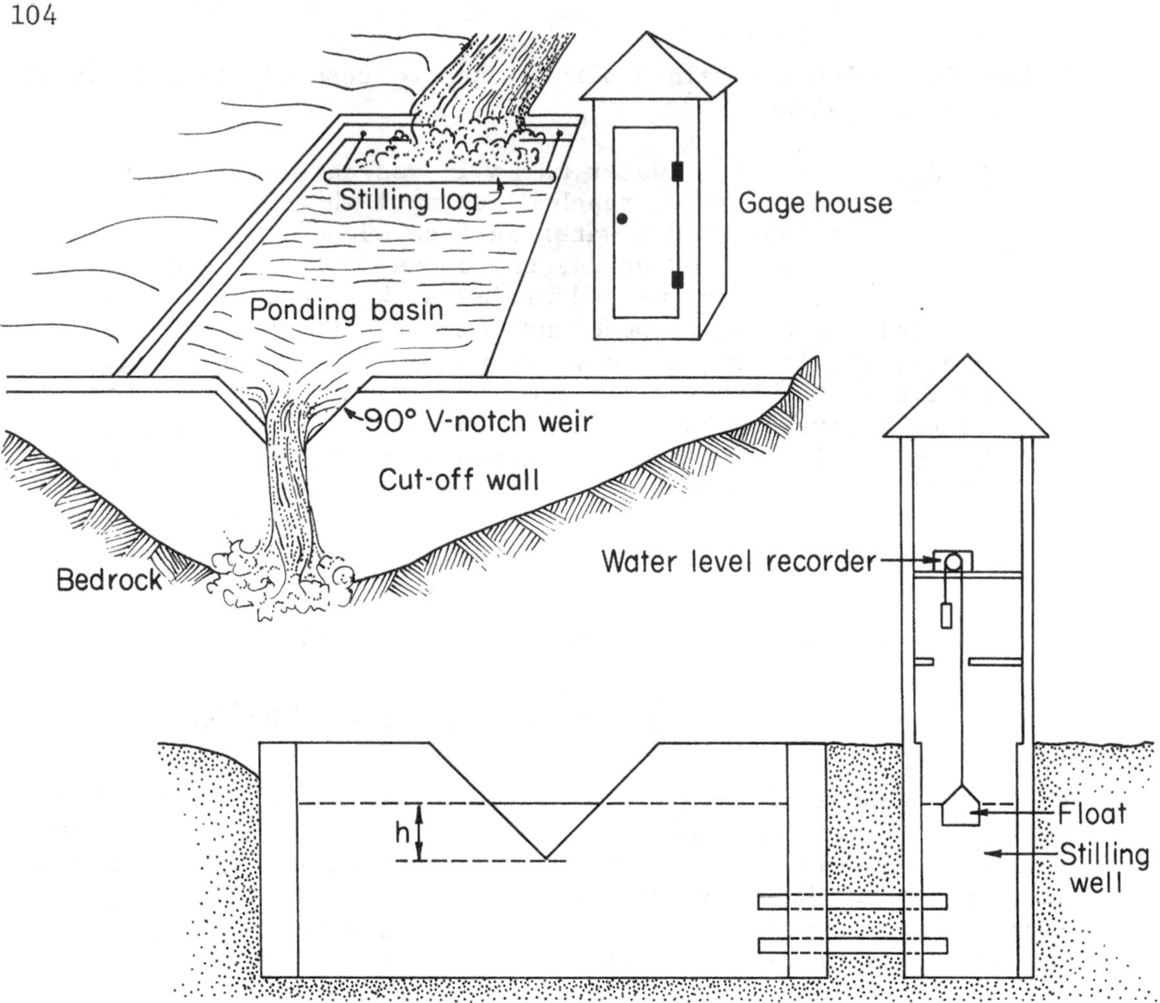

Fig 7-6. Angle and side view of a typical V-notch, sharp-crested weir.

$$A = h^2 \tan(90°/2) \tag{7-4}$$

where h is the height of water above the exact bottom of the notch. The tangent of 45 degrees is equal to one, and so:

$$A = h^2 \tag{7-5}$$

From hydraulic principles, it can be shown that the average velocity $\bar{V}$ through A is a constant C times the square root of h:

$$\bar{V} = C\,h^{1/2} \tag{7-6}$$

Combining Eqs 7-5 and 7-6 the discharge $Q = A\bar{V}$ becomes:

$$Q = h^2\,C\,h^{1/2} = C\,h^{5/2} \qquad \text{UNITS: } L^3T^{-1} \tag{7-7}$$

This is the general form of a rating equation for any 90-degree weir. The coefficient C includes all effects due to the velocity of approach toward the weir blade, the contraction of the water jet as it passes the notch, plus any slight variations in the geometry of the notch. C also carries unit conversions. Many weir and flume types have been rated in hydraulic laboratories; the experimental formula usually given for a properly constructed, 90-degree sharp-crested V-notch weir (Fig 7-6) is:

$$Q = 1.34\ h^{2.48} \qquad 7\text{-}8$$

Q is in m^3/sec and h in m. If ft and ft^3/sec are used, the coefficient C changes from 1.34 to 2.48. Coefficients for other weirs and flumes are listed in hydraulic handbooks (King and Brater 1963).

7. Streamflow records. Charts or tapes from water level recorders are collected and field edited at intervals of once a week to once in six months, depending on the type of recorder employed at the gaging station. Values of head are converted to discharge Q after being visually picked from charts or entered on magnetic tapes by chart readers. Analog-to-digital recorders punch head into paper tape at 5-minute to hourly intervals and all subsequent data reduction and computation can be done by computer. The output is hourly, daily, monthly, seasonal and annual summaries of discharge passing the gaging station. The U.S. Geological Survey publishes monthly summaries as shown in Fig 7-7; the tabular values are mean daily flow rates (ft^3/sec flowing for one day). If the drainage basin is less than 50 square miles (130 km^2) in area, these mean daily flows offer a severely damped version of the hydrograph, insufficient to estimate peak flood discharges and their return periods. However daily, weekly and monthly water yield estimates from such records are unbiased.

Water year. For convenience in reporting hydrologic and meteorologic data, various agencies establish water years that depart from calendar years. The USGS uniformly uses October 1 as the beginning of the water year on the assumption that basin water storage is most often at an annual minimum about that time, and the summer cropping and irrigation season is over. Therefore, tabulated annual water yields of watersheds should vary less because storage tends to assume a threshold value at that time. However, the U.S. Forest Service in the eastern states has found April 1 or May 1 to be a more convenient beginning for some purposes because, under a humid climate, the maximum storage value is more stable from year to year than the minimum in the fall.

Water years like fiscal years can be confusing. It is customary to call the period October 1, 1970 to September 30, 1971, water year 1971. The water year is assigned for tabular purposes to the calendar year during which most of the months fall. The period May 1, 1970 through April 30, 1971, would be called water year 1970.

ALTAMAHA RIVER BASIN

02217500 MIDDLE OCONEE RIVER NEAR ATHENS, GA.

LOCATION.--Lat 33°56'48", long 83°25'22", Clarke County, Hydrologic Unit 03070101, on left bank 0.5 mi (0.8 km) upstream from U.S. Highway 29, 2 mi (3.2 km) west of Athens, and 5 mi (8.0 km) upstream from Barber Creek.
DRAINAGE AREA.--398 mi^2 (1,030 km^2).
PERIOD OF RECORD.--October 1901 to October 1902, January 1929 to March 1932, April 1937 to current year. Monthly discharge only for some periods, published in WSP 1304.
GAGE.--Water-stage recorder. Datum of gage is 555.66 ft (169.37 m) above mean sea level. Oct. 11, 1901 to Oct. 25, 1902, nonrecording gage at site 1 mi (1.6 km) upstream at different datum. Jan. 16, 1929 to Mar. 15, 1932, and Apr. 29, 1937 to Sept. 30, 1940, water-stage recorder at site 4 mi (6.4 km) downstream at different datum.
REMARKS.--Records good. Records of chemical analyses for the water years 1968-74 are published in reports of the Geological Survey.
AVERAGE DISCHARGE.--43 years, 524 ft^3/s (14.8 m^3/s), 17.88 in/yr (454.2 mm/yr).
EXTREMES FOR PERIOD OF RECORD.--Maximum discharge observed, 19,600 ft^3/s (555 m^3/s) Feb. 28, 1902, gage height, 25.5 ft (7.772 m), site and datum then in use; minimum daily, 26 ft^3/s (0.74 m^3/s) Sept. 7, 1957.
EXTREMES FOR CURRENT YEAR.--Peak discharges above base of 3,800 ft^3/s (108 m^3/s) and maximum (*):

Date	Time	Discharge (ft^3/s)	Discharge (m^3/s)	Gage height (ft)	Gage height (m)
Mar. 31	2400	*5050	143	11.92	3.633

Minimum discharge, 109 ft^3/s (3.09 m^3/s) July 23, gage height, 1.11 ft (0.338 m)

DISCHARGE, IN CUBIC FEET PER SECOND, WATER YEAR OCTOBER 1976 TO SEPTEMBER 1977
MEAN VALUES

DAY	OCT	NOV	DEC	JAN	FEB	MAR	APR	MAY	JUN	JUL	AUG	SEP
1	296	353	546	486	550	676	3060	463	351	240	175	125
2	213	288	446	451	494	592	1250	467	328	332	171	119
3	187	267	407	453	495	550	1100	457	308	310	208	116
4	177	257	378	513	503	596	1610	450	292	238	241	119
5	173	246	358	496	507	768	2540	469	284	214	202	137
6	173	238	363	483	503	627	2930	424	274	200	177	153
7	212	235	821	776	490	1010	1510	403	284	190	161	144
8	532	235	846	876	480	1060	1080	393	308	180	170	500
9	666	231	557	950	471	804	924	382	272	170	174	337
10	492	232	468	2610	467	704	832	362	258	163	163	229
11	305	238	432	3210	465	660	776	351	247	157	165	189
12	254	244	495	1400	462	668	732	349	248	148	141	168
13	230	246	854	1040	487	1200	684	343	269	152	130	151
14	229	250	703	985	576	1170	643	332	258	155	138	147
15	214	533	817	1200	539	768	614	325	274	147	126	152
16	204	492	1210	1240	505	656	586	321	254	141	129	288
17	213	359	820	1030	480	600	597	311	297	140	137	1570
18	234	313	622	901	465	555	579	320	320	137	200	1270
19	221	292	543	833	459	530	562	304	288	134	280	556
20	233	283	538	783	455	537	544	300	263	130	200	374
21	319	283	750	762	449	664	520	316	245	121	169	301
22	271	277	629	719	445	1800	509	307	237	116	153	251
23	235	265	538	684	441	1670	537	301	244	111	145	223
24	228	259	500	648	652	944	732	307	241	116	146	210
25	237	257	493	604	904	748	676	317	232	117	156	196
26	368	263	915	553	632	660	562	381	224	123	165	191
27	336	303	941	508	620	618	513	1100	234	140	174	199
28	270	459	675	514	840	583	475	747	223	155	166	210
29	249	1110	597	644	---	596	463	518	210	207	149	198
30	250	945	532	637	---	1470	450	410	207	235	138	185
31	358	---	505	580	---	3560	---	375	---	212	130	---
TOTAL	8579	10253	19299	27569	14836	28044	28590	12605	7974	5331	5179	9008
MEAN	277	342	623	889	530	905	953	407	266	172	167	300
MAX	666	1110	1210	3210	904	3560	3060	1100	351	332	280	1570
MIN	173	231	358	451	441	530	450	300	207	111	126	116
CFSM	.70	.86	1.57	2.23	1.33	2.27	2.39	1.02	.67	.43	.42	.75
IN.	.80	.96	1.80	2.58	1.39	2.62	2.67	1.18	.75	.50	.48	.84

CAL YR 1976 TOTAL 247588 MEAN 676 MAX 11600 MIN 145 CFSM 1.70 IN 23.14
WTR YR 1977 TOTAL 177267 MEAN 486 MAX 3560 MIN 111 CFSM 1.22 IN 16.57

Fig 7-7. Sample page from Water Resources Data for Georgia, Water Year 1977, U.S. Geological Survey Water-data Report GA-77-1.

B. Runoff terminology. The early job of the hydrologic engineer had more to do with flood prediction and design of structures on major streams than with understanding and explaining the effects of climate, soils, vegetation and land use in the source areas. Recently, hydrologists have devoted more attention to source area processes because problems with water yield, erosion, pollution, mineral cycling and the continuing controversy over the source and cause of floods in the headwaters were intensifying. Large public programs are involved in water resource development, totaling billions of dollars a year; some of this money may be misdirected because of incomplete information about the land phase of the hydrologic cycle. Classical hydrological techniques, developed mostly for downstream use, dominated the terminology of the runoff process and tended to delay communication between the resource planner, the hydrologist and the land manager. Glossaries of hydraulics and hydrology (e.g., Nomenclature for Hydraulics, American Society of Civil Engineers 1962) contain a confusing array of multiple definitions for commonly used terms. The land manager needs terms that are clearly defined and consistent.

The first-order perennial stream basin, which may be visualized as occupying all land areas except the flood plains of major rivers, is the domain within which the following terminology applies. Runoff is a summary term used to refer to the various processes that ultimately produce streamflow. The word "runoff" by itself does not convey any idea of the source or timing of the water yielded by a basin, and therefore should not be used (as it all too often is) as a substitute for more precise terms describing the components of streamflow.

1. Flow components classified. Obviously all precipitation does not immediately flow out of a basin. Some water flows out fairly quickly, some is stored temporarily and some never flows out via the stream channels because it evaporates back to the atmosphere or percolates to deep ground water aquifers. The following terms are in common usage by watershed managers to classify and describe the complex runoff process.

Channel precipitation (C_p) is that part of streamflow derived from net precipitation falling directly into the flowing stream. The area receiving channel precipitation is generally about one percent of the total basin area in the East, since this is the area occupied by the perennial channel system. However, during a prolonged storm the area receiving channel precipitation will increase to several percent of the basin area due to expansion of the stream into intermittent and ephemeral channels.

Overland flow (R_s) is that part of streamflow derived from net precipitation which fails to infiltrate the mineral soil surface and runs over the surface of the soil to the nearest stream channel without infiltrating at any point.

Surface runoff is often used synonymously with overland flow, but it may mean any water measured in surface streams of a region. An exact definition of overland flow is critical to the understanding of source area hydrology.

Surface stormflow ($C_p + R_s$) is the sum of the above two components of flow. The important distinction is that surface stormflow has not infiltrated the mineral soil surface, while subsurface stormflow has infiltrated.

Subsurface stormflow (R_i) refers to that part of streamflow which derives from subsurface sources but arrives at the stream channel so quickly that it becomes part of the storm hydrograph produced directly by a given rainstorm. Since classification of stormflow is arbitrary, depending on the judgement of the hydrologist, and since subsurface flow is not visible to the eye, there is much uncertainty as to the "separation" of subsurface stormflow from baseflow. But there is little doubt that the largest component of stormflow from forests and most wildlands begins as subsurface flow. Sometimes the term interflow is used synonymously with subsurface stormflow, but other times it is used to refer to any non-vertical subsurface flow above the water table.

Stormflow (direct runoff, Q_s) is the sum of surface and subsurface stormflow ($Q_s = C_p + R_s + R_i$) and is the term most often used by hydrologists in describing the flood-producing characteristics of watersheds. In classical hydrology, direct runoff was assumed to be entirely overland flow (Horton 1945, Sherman 1932; see Chow, 1964).

Baseflow (ground water outflow, R_g) is normally thought to be the sole component of streamflow between storm or snowmelt periods, and thus baseflow is presumably the oldest water to be yielded by the basin. Generally, baseflow is defined as outflow from extensive ground water aquifers which are recharged by water percolating vertically through the soil mantle to the water table (refer to Fig 5-7). But baseflow is also sustained by slow drainage of unsaturated soil in the zone of aeration, particularly in steep areas. In upland streams draining good forest land, about 85% of the total streamflow may be termed baseflow. In the East as a whole, about 70% of total streamflow is baseflow, and only 30% is stormflow.

Streamflow (Q) is the flow of water past any point in a natural channel above the bottoms and sides of the channel. Quantitatively streamflow is a rate of discharge measured at a gaging station (Sec A2), including any underflow forced to surface at the gaging site by a cut-off wall. Streamflow is the sum of the following components already defined above:

$$Q = C_p + R_s + R_i + R_g \qquad 7\text{-}9$$

Deep seepage (basin leakage L) is the loss of water from a drainage basin by deep pathways that do not discharge into the channel above a gaging station or design site. Loss may be downward into regional groundwater aquifers or into caverns and underground streams. Another pathway for leaks occurs laterally under the surface water divide (Chap 3). Deep seepage may be either a gain or a loss to basin streamflow, depending on where the water surfaces.

Underflow (U) refers to ungaged water moving past a stream channel section in valley sediments or colluvial material. In some terranes underflow may be a large portion of total water yield of basins because controlled sections cannot be anchored to bedrock.

Water yield (WY) is a drainage basin's total yield of liquid water during some period of time. Water yield is equal to the difference between gross precipitation (P_g) and evapotranspiration (E_t), corrected for change in storage:

$$WY = P_g - E_t - \Delta S \qquad 7\text{-}10$$

In terms of flow components, WY may also be defined:

$$WY = Q + U \pm L \qquad 7\text{-}11$$

In experimental watersheds the hydrologist always seeks basins where U and L are small, constant or negligible.

There are only 11 internally consistent terms essential to definition of small watershed hydrographs. The land manager will come across many additional terms to describe sources and pathways of flow, but they are neither rigorously nor consistently defined. Relate these 11 to Fig 7-1.

Defined from the hydrograph: Streamflow, peakflow, antecedent (initial) flow, time of rise.

Classified from the hydrograph (classification is arbitrary): Stormflow (direct runoff), stormflow duration.

Defined by process description: Surface stormflow, overland flow, channel precipitation. (While definable, these 3 are not separable on natural basins except by some arbitrary rule or arbitrary instrument.)

Classified by subtraction: Baseflow = Streamflow - Stormflow; Subsurface Stormflow = Stormflow - Surface Stormflow.

No graphical or mathematical operation performed on a hydrograph will reveal the source or pathway of stormflow. The changing rate of stream discharge is the net effect of accretion and depletion of many components of storage in the basin. Different combinations of sources and pathways can lead to quite similar hydrographs.

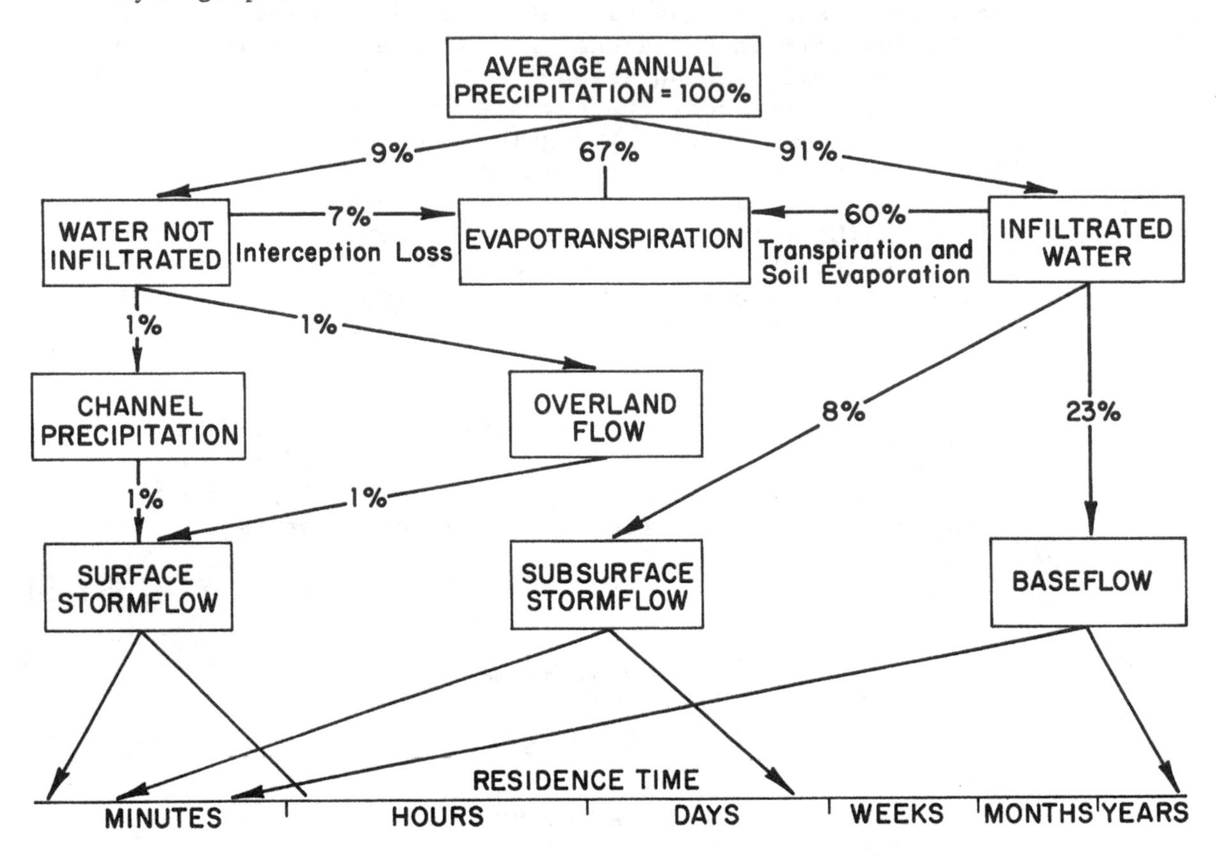

Fig 7-8. In the humid areas of eastern U.S., annual precipitation is disposed of approximately as shown by the percentages in this diagram. Surface stormflow leaves the small basin within minutes, whereas some baseflow may be retained within the soil mantle for years.

2. Storage components classified. Even the most rapidly delivered water from a basin is delayed for some period of time, so the hydrologist finds it necessary to classify components of storage along the pathway from precipitation to streamflow. Fig 7-8 suggests the relation between the residence time (the "age" of water) and flow components (based on records in eastern U.S.).

Interception storage is water or snow held on the aboveground surfaces of vegetation. The amount held by vegetation varies from about 0.5 mm in cropland to about 1.5 mm in thick coniferous forest. Interception storage usually becomes interception loss (Chap 6), but interception has little effect on stormflow except in small storms, when channel precipitation may be reduced 50% by vegetation hanging over the stream. In major storms, interception storage is a negligible amount compared to the total precipitation over those areas actually producing channel precipitation or overland flow. Elsewhere all rain would infiltrate if not stored on vegetation.

Snow storage, or water held in the snowpack, obviously does not contribute to any form of streamflow until the weather moderates and the snow melts. Snow storage and soil freezing can stop streamflow altogether in alpine and polar regions.

Surface retention storage is the thin film of water which must wet the soil surface before flow can begin. Seldom more than 0.5 mm, surface retention storage is an appreciable abstraction from stormflow only when infiltration is greatly impeded and overland flow is the dominant source of stormflow. Where infiltration exceeds rainfall rate, surface retention storage need not be accounted for as an abstraction from stormflow.

Surface detention storage is rainwater or snowmelt detained temporarily on the surface by the resistance of surface irregularities to flow downslope. Surface detention storage offers a large opportunity for infiltration before the water enters the stream. Detention and retention storages are artifically classified, thus never precisely separable.

Forest floor retention storage is the throughfall and stemflow that is retained by the intercepting litter, fermentation and humus layers. Forest floors vary widely in depth and type, so the amount of water retained varies greatly. Nearly all the water retained by litter, and some of that retained by humus and fermentation layers, will become forest floor interception loss (I_f, Chap 6) between storms. Roots have access to the F layer, particularly in mull humus, so the distinction between forest floor retention and soil water is not precise.

Forest floor detention storage is the throughfall and stemflow detained temporarily in the litter, fermentation and humus layers. Similar to surface detention storage, but larger in quantity, forest floor detention plays a large role in reducing stormflow during extreme events because it holds short bursts of rain for later infiltration. Even shallow forest floors, such as a scattering of pine straw under young plantations, detain water long enough to take care of all but the most intense bursts of rain. This benefit is in addition to the main hydrologic role of the forest floor, which is to absorb the impact of raindrops and to deliver clean water to the mineral soil, where, with some exceptions, the infiltration capacity is adequate to handle it. There is downslope shifting of forest floor detention storage if rainfall continues; this has been called "litter flow" and "shingle flow." When shingle flow occurs along expanding channel networks, the shingled water immediately becomes part of surface stormflow. This contribution to storm water is best thought of as part of channel precipitation. "Litter flow," defined as a basin-wide surface flow of water, is sometimes offered as an erroneous explanation of stormflow from forest land.

<u>Depression storage</u> is the water collected in pools and depressions shortly after rainfall or snowmelt. Depression storage is an appreciable abstraction from stormflow in limestone sinkhole areas (karst topograph), where it can greatly reduce direct runoff by reserving surface water for later infiltration.

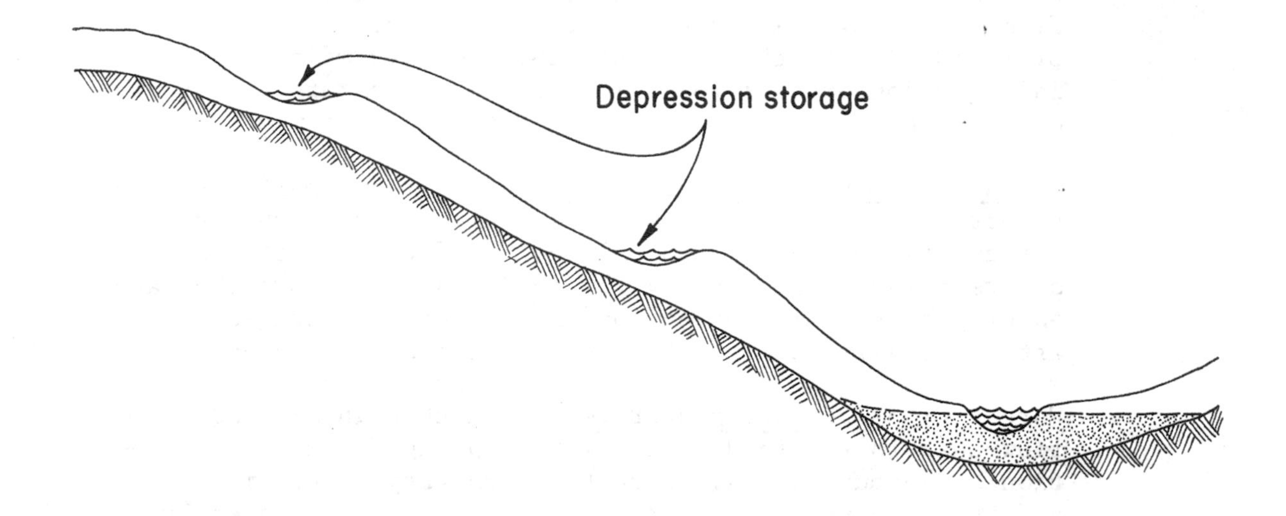

Another example of depression storage is often seen behind contour terraces following storms. Litter dams in the forest floor sometimes play the role of depression storage.

<u>Soil moisture storage</u> (Chap 5) may be separated into water detained in the weathered mantle for only short periods of time (detention) after a storm and that retained for longer periods (retention). When the zone of aeration is deep, dynamic storage in the soil mantle plays a large role in the timing and quantity of both stormflow and baseflow.

<u>Ground water storage</u> (Chap 5) may remain within the basin for years, but in saturated zones along stream channels ground water may quickly discharge as stormflow. The role of ground water in stormflow is not yet well defined.

<u>Channel storage</u> is the water in the stream or river channel at any given time. Channel storage varies greatly during and following a storm; its effect on the downstream river hydrograph is dominant. Flooded upstream bottomlands temporarily hold water back from adding to flood peak discharges further downstream. Channel and flood plain storage produce special problems in hydrologic prediction, involving techniques of flood routing. Land use and vegetal cover have limited influence on the behavior of storm water once it is in the channel, except for those activities which affect the stability of channel banks. Examples of the latter are channel dredging, snagging, flood plain alterations and the growth or cutting of vegetal stands along the banks.

Infiltration was once called a "storage loss" from the storm hydrograph because it was wrongly thought that infiltrated water moved too slowly underground to contribute stormflow. The study of discharge from small plots led to this misconception and the quick appearance of surface water in streets, lawns and compacted fields after rain encouraged laymen to accept it for many years. Some engineering hydrology textbooks still label all stormflow (direct runoff) as overland flow or surface runoff. Only on totally impervious basins will stormflow be equal to overland flow; no forests or wildlands fit this category.

In summary, two major sets of factors control the hydrograph of discharge from basins: Watershed factors (Chap 3) and weather factors (Chap 4). In brief, weather factors are:

Amount of rainfall (total cm per storm)
Intensity of rainfall (cm/hour)
Duration of rainfall (hours, days, even weeks)
Distribution of rainfall or snowmelt over the basin (high and low elevation)
Temperature (frozen soil, snow and ice melt).

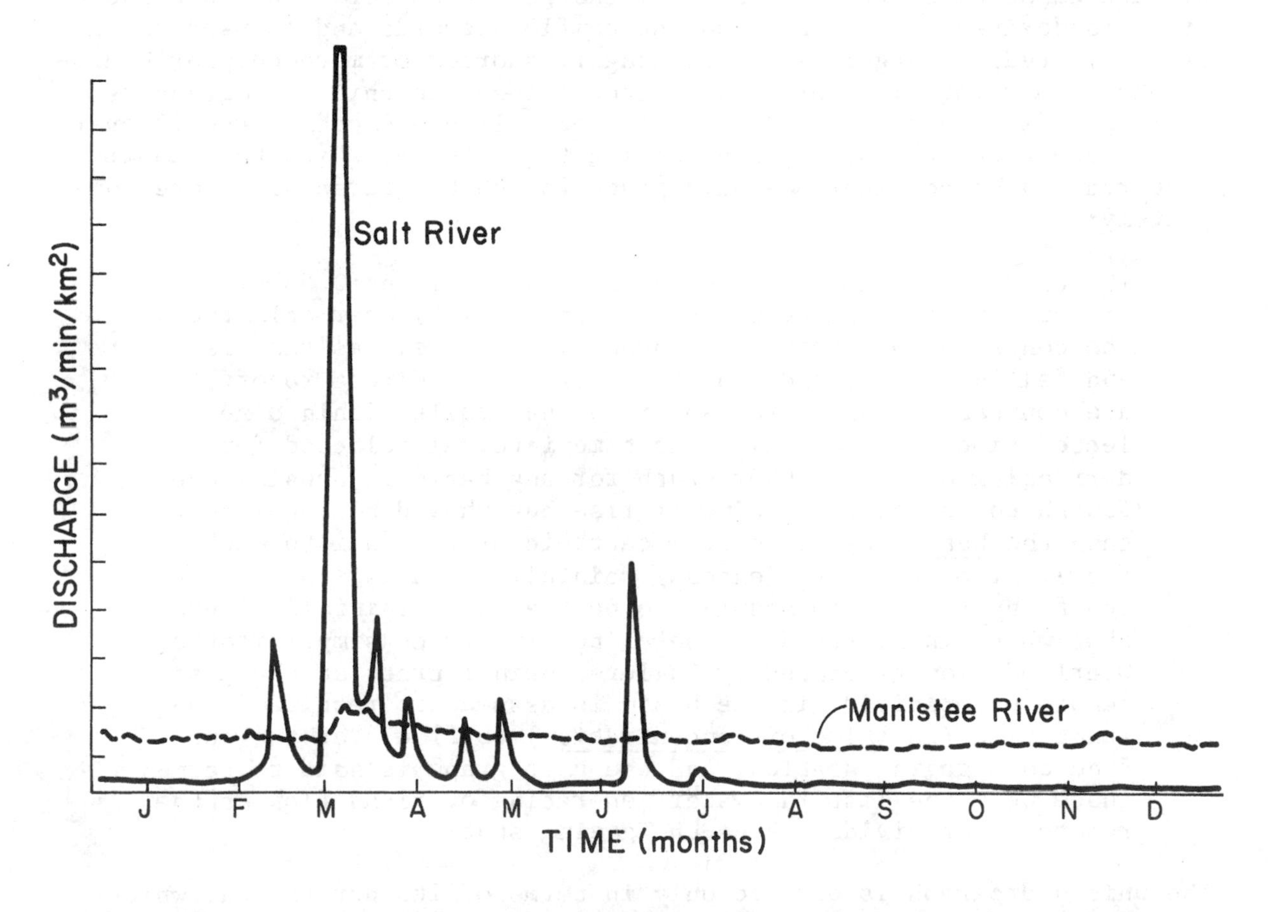

The annual hydrographs for two streams under similar precipitation regimes in central Michigan are diagrammed above. The Manistee River hydrograph varies little throughout the year, indicating excellent storage in deep, permeable soils with good infiltration. The Salt

River hydrograph shows a flashy response to precipitation and snowmelt, revealing poor storage caused by impermeable or shallow soil layers. The latter conditions are associated with the clayey glacial tills of the area.

C. The unit hydrograph. By 1932 a sufficient number of years of streamflow and precipitation record had accumulated to cause questions to be raised about the usefulness of these data in prediction and design work. At first every rainstorm and its storm hydrograph appeared to be unique and anyone designing a dam or bridge was forced to study local records and to form his own opinion about the recurrence interval of peak discharges, the volumes of water to be stored or passed in a major flood and the routing coefficients to use in predicting a moving flood. About this time L. K. Sherman published a method for predicting and routing stormflow that became famous as the "unit hydrograph method." Sherman noted in the accumulated record that bursts of any amount of rainfall within a specified time interval over a particular basin tended to produce storm hydrographs of similar shape and duration. He therefore decided that there was a "unit graph" for each basin that represented the average capacity of that basin to discharge stormwater. The unit hydrograph is derived once from actual streamflow records and is used thereafter to predict larger, smaller, longer, shorter or more complex hydrographs from rainfall records. Sherman defined the unit hydrograph as follows: "If a one-day rainfall produces a 1-inch depth of runoff over a drainage area, the hydrograph showing the rates at which the runoff occurred can be considered a unit graph for that watershed." More completely:

> The unit hydrograph represents exactly 1.0 inch (2.54 cm) of direct runoff (assumed to be overland flow by Sherman) from the contributing basin. The peak flow, as well as the rising and falling limbs, and thus the duration of direct runoff, are constant in shape for any rain that falls within a selected time interval (t). The time interval selected for derivation of a unit hydrograph for any basin is usually one-fourth to one-half the time of rise but should not be longer than the basin lag (time from centroid of rainfall to peak flow). The effective (excess) rainfall producing the direct runoff hydrograph is assumed to be the total rainfall minus that which infiltrated, a useful but incorrect simplification. Overland flow generated by failure to infiltrate at the most remote water divide in the basin is assumed to require a constant time (the time of concentration) to flow over the surface to a gaging station, and the unit graph is said to terminate when the last rainwater generating overland flow at the remote water divide passes the gaging station.

The unit hydrograph is defined only in terms of its derivation, which is based upon the assumption that the storm hydrograph is produced by overland flow and that the entire drainage basin contributes water to stormflow proportional to area. As described, the unit hydrograph method strictly applies only to areas where uniform rates of overland

BASIN AREA = 100 hectares

UNIT GRAPH = 100 m^3 X 10^2

STORMFLOW = 150 m^3 X 10^2

MULTIPLIER = 100/150

= 0.667

Clock Time (from-to)	Actual Stream-flow	Base-flow	Storm-flow	Unit Graph Ordinate
hr	-----	m^3/hr X 10^2	-----	
7 - 8	1.0	1.0	0	0
8 - 9	3.3	1.1	2.2	1.5
9 -10	13.2	1.2	12.0	8.0
10 -11	29.7	1.3	28.4	18.9
11 -12	35.6	1.4	34.2	22.8
12 -13	26.8	1.5	25.3	16.9
13 -14	20.2	1.6	18.6	12.4
14 -15	14.9	1.7	13.2	8.8
15 -16	10.5	1.8	8.7	5.8
16 -17	7.0	1.9	5.1	3.4
17 -18	4.3	2.0	2.3	1.5
18 -19	2.0	-	0	0
			150.0	100.0

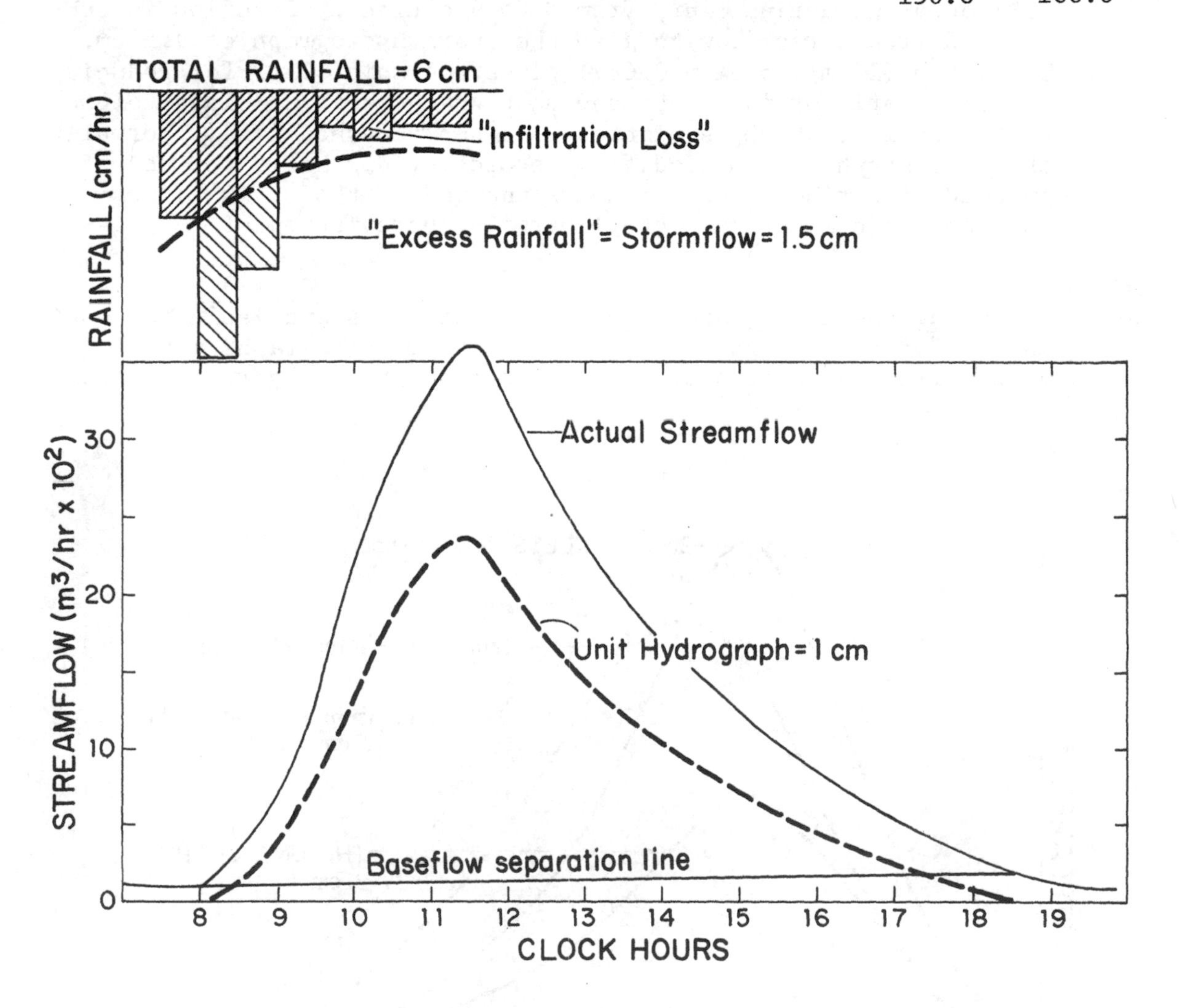

Fig 7-9. See text for explanation of unit hydrograph derivation.

flow are generated from the entire basin above the gaging station; parking lots, airfields, compacted bare soil and some naturally impervious areas (rock outcrops) may fit this condition. Only small areas within forest, range and wildlands conform to this concept of flow production, and there are many exceptions in agricultural land as well. Consequently an elaborate array of adjustments and alternate methods have developed over the years to adapt the unit hydrograph technique to watersheds where overland flow is not the predominant source of stormflow.

The unit hydrograph method has its main advantage in the simplicity of its basic concept and its usefulness in engineering design work.

Example: Using 1 cm of stormflow (excess rainfall) for the unit graph method, Fig 7-9 shows a hydrograph for a 100-ha basin, an actual record in which 1.5 cm of stormflow was discharged from 6 cm of rainfall in 4 hrs. Applying the theory of the infiltration capacity curve (Chap 5, Sec G1), we see that excess rainfall occurred during 1 hr, from 8 to 9 o'clock. Baseflow is subtracted from streamflow to give the storm hydrograph of 1.5 cm. The ratio (1 cm/1.5 cm = 0.667) gives the multiplier for reducing all the hourly ordinates to the unit graph ordinate (last column of the table), which is plotted as a heavy dashed line. Normally the unit graph is computed from records of discharge per second instead of per hour; use of hours and the unit $m^3 \times 10^2$ (one hundred cubic meters of water) permits abbreviation of computations.

1. Using the unit hydrograph. If a method is available for estimating "effective" rainfall, complex storm hydrographs may be predicted from the unit hydrograph. To predict the hydrograph from

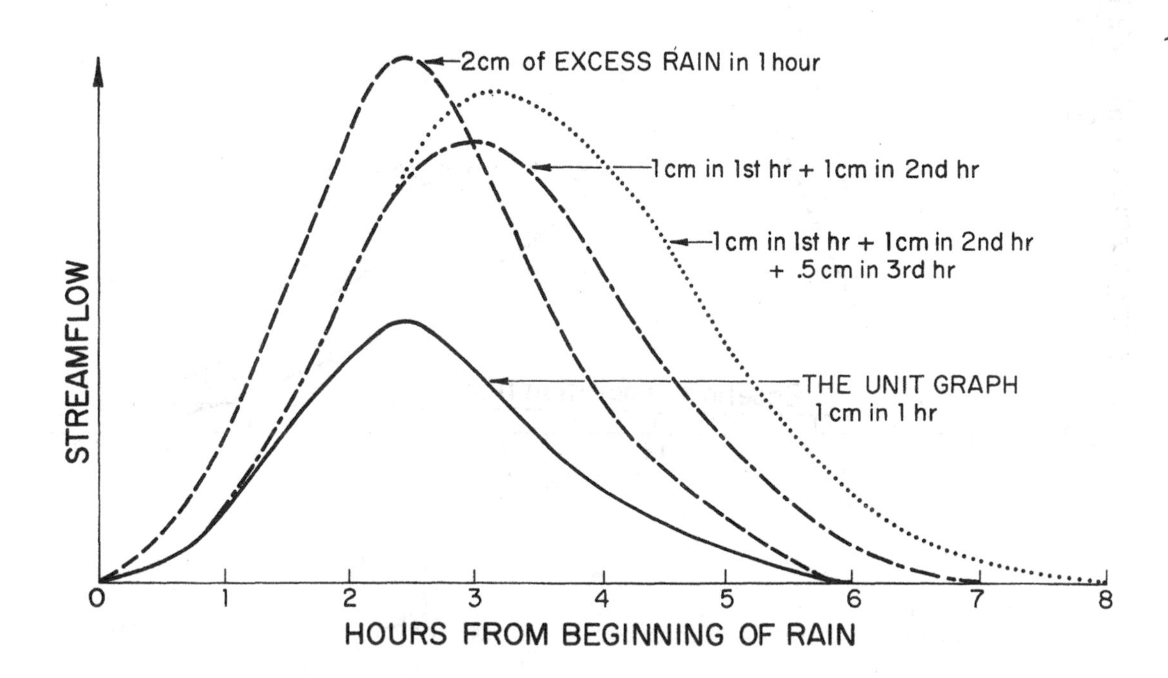

a storm delivering 2 cm of effective rainfall, we superimpose the unit graph on itself; that is, add the hourly ordinates to give twice the peak and twice the flow throughout, while retaining the same duration. If we want the hydrograph of 1 cm of effective rainfall followed by a second cm in the next hour, we lag the unit graph one hour and add as before; stormflow duration has increased by one hour. If 0.5 cm of effective rain occurs in the third hour, we divide all ordinates of the unit hydrograph by two, lag another hour and add again. Complex hydrographs are thus predicted but their accuracy depends on all the assumptions of the method as well as on how "effective rainfall" is estimated.

The unit hydrograph method is standard in engineering hydrology and is almost universally used in design of bridges, dams, levees, flood ways, storm sewers, as well as in flood prediction and flood plain mapping. The techniques of flood routing are also required since the stream is seldom gaged at the reach of particular interest.

2. Flood routing is a technique used to determine the effect of channel or reservoir storage on the shape and movement of the storm hydrograph. As stormflow moves downstream its time base (stormflow duration) lengthens and the peak is damped. In elementary form, flood routing is easy to understand but application requires extensive knowledge of channel or reservoir characteristics. Consider the 1-ha pond shown in Fig 7-10; assume that the pond is full to the pipe but no water is entering or leaving. We wish to route through the pond a 1-cm stormflow from the 100-ha basin above. The input hydrograph, therefore, is the unit graph in Fig 7-9. The routing equation is a statement of the conservation of matter:

$$\text{Input} = \text{Output} + \text{Change in Storage}$$

Input (I) is the average inflow rate at the head of the pond, output (0) is the average rate of flow through the overflow pipe and change in storage (ΔS) is the gain or loss of water to the pond. For this simple example the routing interval is set at 1 hr, and the subscripts 1 and 2 indicate the beginning and end of the routing interval. Therefore, the average balance during the hour is:

$$\frac{I_1 + I_2}{2} = \frac{0_1 + 0_2}{2} + (S_2 - S_1) \qquad 7\text{-}12$$

Just as the inflow hydrograph arrives at the pond, we know that

$$I_1 = 0_1 = S_1 = \text{zero}$$

and also that I_2 will be the rate of flow from the unit graph at the end of the first hour ($4\ m^3/hr \times 10^2$, at 9 o'clock, Fig 7-9).

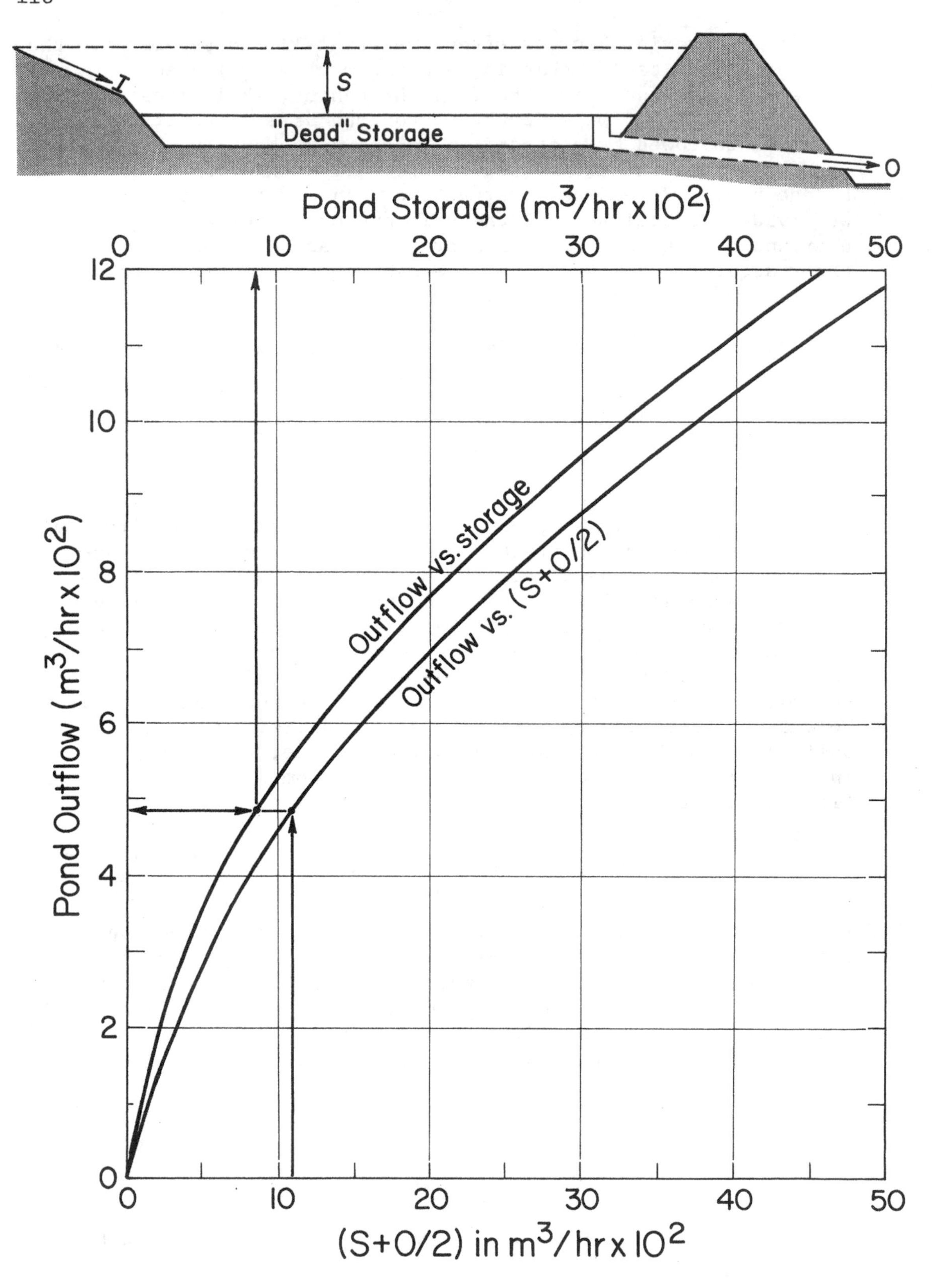

Fig 7-10. Diagram of a 1-ha pond and the outflow-versus-storage curve developed from the pond's dimensions and the hydraulics of the outflow pipe. See text and Fig 7-11 for instructions on how to route stormflow through the pond.

Rearrange Eq 7-12 to put all knowns on the left and unknowns on the right:

$$I_1/2 + I_2/2 + S_1 - O_1/2 = S_2 + O_2/2 \qquad 7\text{-}13$$

This is the routing equation in its usable form; it appears again in the computation table, Fig 7-11. However, an equation in two unknowns (S_2 and O_2) cannot be solved without another equation, one that relates S and O in the pond. The storage-outflow equation is unique for each pond because outflow through the pipe is highly sensitive to water depth over the pipe, and each increment in head represents quite different volumes stored in the pond. Abbreviating these details, we provide the storage-outflow curve in Fig 7-10, assumed to be surveyed at the site.

Using the outflow-storage and outflow -(S + O/2) curves as a nomograph, each line in the table in Fig 7-11 is solved to obtain the pond outflow at the end of each routing hour. The solution at 8 o'clock gives 2 $m^3/hr \times 10^2$ for the value (S + O/2). Enter the graph at the bottom and read pond outflow from the curve labeled (S + O/2). The outflow is 1.4 $m^3/hr \times 10^2$ which is entered in the last column opposite 9 o'clock. The value 1.4 is also O_1 for the next iteration, so 1.4/2 is entered in Column $O_1/2$ opposite 9 o'clock. Knowing the outflow at 9 o'clock allows us to update the storage in the pond: Read pond storage on the upper scale ($m^3 \times 10^2$), using the outflow-storage curve. We find that the storage equivalent to 1.4 outflow is 1.0 $m^3 \times 10^2$. Enter it at 9 o'clock under S_1, which is the pond storage at the beginning of the second routing hour. Add and subtract as indicated and compute a new (S + O/2). Continue until virtually all the water is routed (outflow asymptotes to zero but we cannot read the nomograph that precisely). Routed peakflow is halved and stormflow duration is extended about 6 hours.

Flood routing is routine in design of spillways and all other streamflow control structures.

3. Runoff curves. The unit graph method assumes as a matter of definition that the volume of excess rain equals the volume of stormflow. Estimation of excess rain is based on theoretical infiltration capacity curves (Chap 5), which are interpreted from soil type maps. Most soil types have been classified into 4 groups:

Hydrologic Soil Group	Infiltration Capacity
A	Fast
B	Moderate
C	Slow
D	Very slow

O'CLOCK	INFLOW $m^3/hr \times 10^2$	$I_1/2$	+ $I_2/2$	+ S_1	− $O_1/2$	= $(S_2 + O_2/2)$	OUTFLOW $m^3/hr \times 10^2$
8	0	0	+ 2.0	+ 0	− 0	= 2.0 →	0
9	4.0	2.0	6.8	1.0	.7	9.1	→ 1.4
10	13.5	6.8	11.1	6.9	2.2	22.6	4.3
11	22.3	11.1	10.2	18.9	3.8	36.4	7.4
12	20.3	10.2	7.0	31.6	4.9	43.9	9.8
13	14.0	7.0	5.2	38.5	5.4	45.3	10.8
14	10.4	5.2	3.7	39.8	5.6	43.1	11.6
15	7.3	3.7	2.2	37.5	5.4	38.0	10.7
16	4.5	2.2	1.1	32.9	5.1	31.1	10.0
17	2.3	1.1	.7	26.9	4.5	24.2	9.0
18	1.4	.7	0	20.4	3.9	17.2	7.7
19	0	0	0	14.0	3.2	10.8	6.4
20	0	0	0	8.5	2.4	6.1	4.9
21	0	0	0	4.4	1.6	2.8	3.3
22	0	0	0	1.8	.9	.9	1.8
23	0	0	0	.6	.3	.3	.7
24	0	0	0	.1	0	.1	.1
1	0	0	0	0	0	0	(Trace)
SUM	100.0						99.9

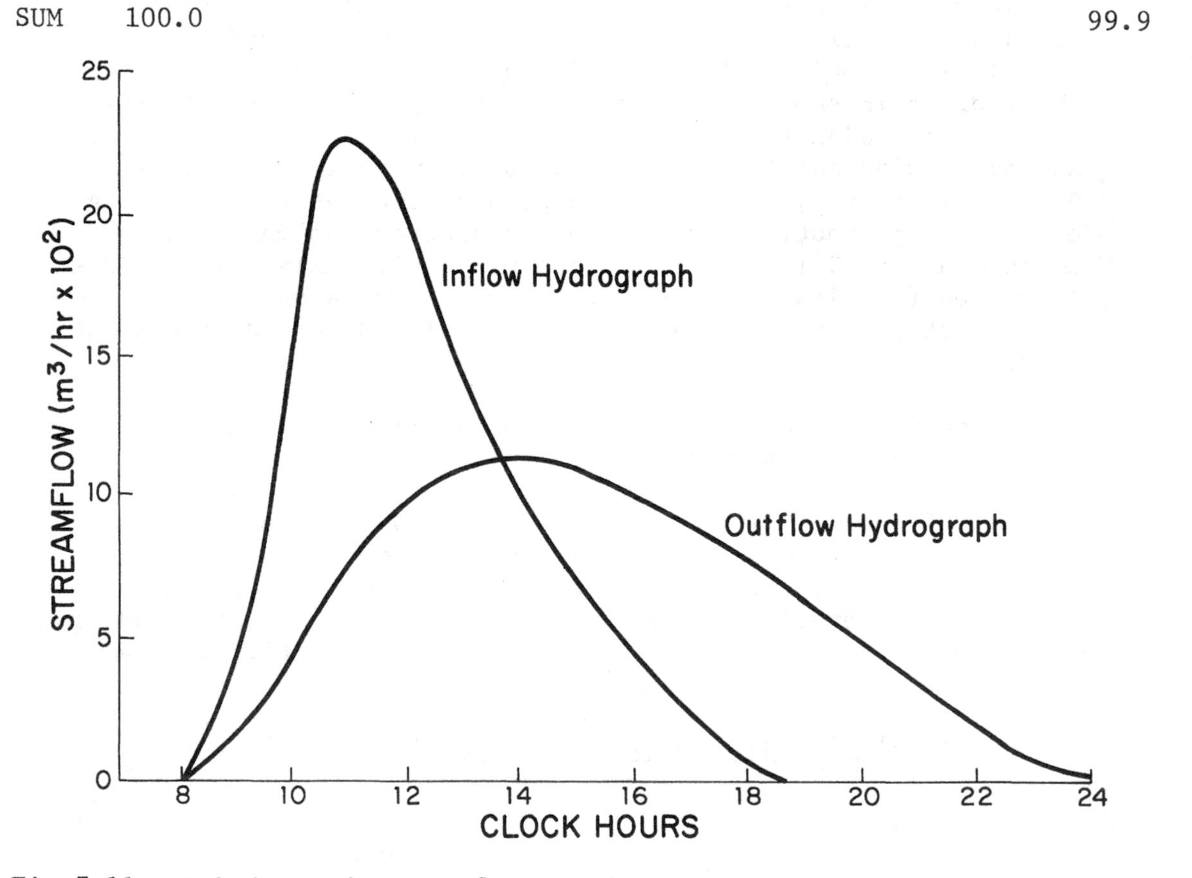

Fig 7-11. Tabular solution of Eq 7-13 to route the unit stormflow in Fig 7-9 through the pond in Fig 7-10. The exercise shows how flood flows are damped and lagged by storage. See text for explanation of the tabular solution.

The hydrologic soil group is further modified by land use (fallow, crop, pasture, woodland); conservation practice (clean-rowed, contoured, terraced); and hydrologic condition (judged in the field as good, fair or poor). Each combination of soil group, land use, hydrologic condition and soil practice is associated with a runoff curve that expresses stormflow (Q) in relation to storm rainfall (P).

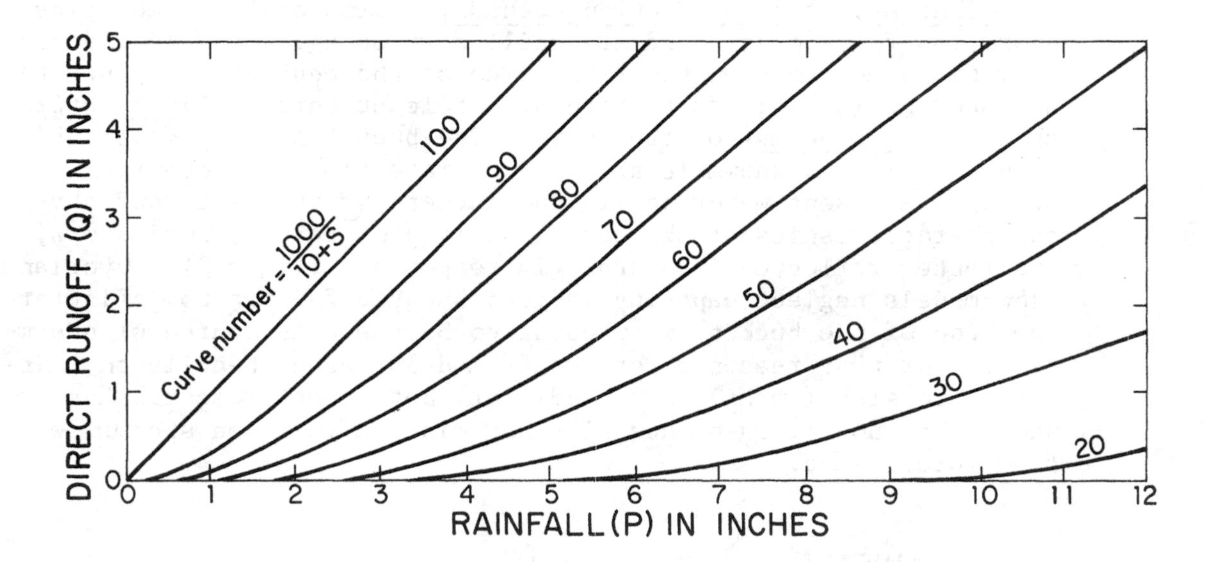

The runoff curve, like the unit hydrograph method, is based on the assumption that stormflow is overland flow and that the entire basin or soil type generates stormflow uniformly whenever the rainfall rate exceeds the theoretical infiltration capacity of the soil surface. Early observation of field plots (5 ha or less) led to a preliminary assumption that about 0.2 of the plot's "potential infiltration" (S) was withheld from any rainstorm as an "initial abstraction" before any "runoff" was generated. On this assumption, a general function between stormflow and storm rainfall was devised (SCS 1972):

$$Q = \frac{(P - .2S)^2}{P + .8S} \qquad 7\text{-}15$$

Q is stormflow, P is storm rainfall and S is "potential infiltration," all in inches per storm. S is also referred to as "the maximum possible difference between rainfall and runoff" and "maximum possible storage." In fact, it is a complex transformation of P and Q derived by solving Eq 7-15:

$$S = 5[P + 2Q - (4Q^2 + 5PQ)^{1/2}]$$

These formulae have no physical meaning, except when S is zero (no storage) and therefore Q equals P. However, Eq 7-15 roughly approximated stormflow in the agricultural land of the prairie states, where the method was devised. Though widely used, few hydrologists are satisfied with the accuracy attained by the runoff curve method. Large stormflows and peakflows from forests and wildlands will generally be overestimated by a factor of 2 or more, while smaller flows will be estimated as zero because of false assumptions about "initial abstractions" (Hewlett et al. 1977).

4. Other stormflow prediction methods. Formulae for predicting stormflow from forest land are still developing. Eq 7-15 contains only two predictors on the right side of the equation: Input (P) and storage (S). In fact there are at least three: Input, storage and storage capacity of the basin. The bucket-analogy makes this clear. Rain and snowmelt are inputs, retention and detention storage represent water now in the bucket and the depth and physical characteristics of the earth mantle are capacity indicators, the latter reflected in hydrologic response R (Fig 7-2). Overland flow models neglect capacity indices because failure to infiltrate the "top of the bucket" is thought to be the main source of stormflow. For that reason older runoff models leaned heavily on rainfall intensity (cm/hr) as a predictor, but recent research has shown that hourly rain intensity has minor effects on stormflow from forest land.

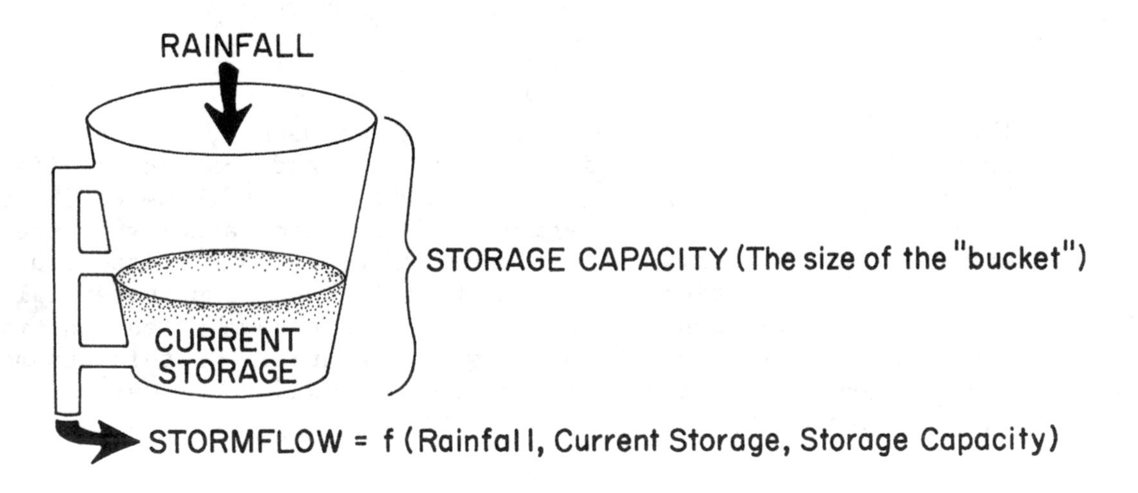

5. An R-index formula. The following formulae (Hewlett et al. 1977) will give planning estimates of stormflow (Q_s) and peakflow (Q_p) from forest land in eastern U.S., using average hydrologic response R from the map in Fig 7-2.

$$Q_s = 0.35 \; R \; P_g^{1.5} \qquad 7\text{-}16$$

and

$$Q_p = 3.4 \; R \; P_g^{1.6} \qquad 7\text{-}17$$

Q_s is in cm, Q_p is in $m^3/min/km^2$, R is a dimensionless fraction and P_g is gross storm rainfall in cm. The coefficients include unit corrections. Answers calculated must be seasonally adjusted for current storage by these multipliers:

AUG	SEP	OCT	NOV	DEC	JAN	FEB	MAR	APR	MAY	JUN	JUL
1.0	1.0	1.1	1.2	1.3	1.4	1.5	1.5	1.4	1.3	1.2	1.1

Formula 7-16 is subject to the constraint that increments in Q_s cannot exceed increments in P_g. By differentiation of 7-16, we show that when P_g reaches $3.6/R^2$, all <u>further</u> increases in storm rainfall must become stormflow. Values for $3.6/R^2$ are too large to worry about unless R exceeds 0.35 or P_g exceeds 30 cm:

When R =	.10	.15	.20	.25	.30	.35	.40	.45	.50
P_g limit (cm) =	360	160	90	58	40	30	23	18	14

For example, if R = 0.45, all storm rainfall above 18 cm goes directly into stormflow because no more basin storage is available (the "bucket" is full).

<u>Example</u>: How many m^3 of water would be added to a 1-ha pond below a 100-ha basin in the Southeastern Piedmont if a 25-cm rainstorm occurred during March? The response map Fig 7-2 suggests an average R = 0.16. (P_g did not exceed the constraint.)

$$Q_s = .35(.16)(25^{1.5}) = 7 \text{ cm (unadjusted)}$$

Adjust for March response:

$$1.5\ (7\ \text{cm}) = 10.5 \text{ cm of stormflow;}$$

and

$$\frac{(10.5\text{cm})}{(100\text{cm/m})}\ (100\text{ha})(10000\text{m/ha}) = 105000\text{m}^3 \text{ added to pond.}$$

What was the peakflow in $m^3/min/km^2$?

$$Q_p = 3.4\ (.16)(25^{1.6}) = 93.8\ m^3/min/km^2$$

Since 100 ha equals 1 km^2, peak discharge was 93.8 m^3/min.

D. <u>The runoff process clarified</u>. The land manager can no longer be satisfied with the simplistic concept of the runoff process that served as a base for the unit hydrograph and runoff curve techniques. To explain and predict the behavior of water, minerals, nutrients, pollutants and soil erosion on the land, the manager needs an understanding of source area processes that the Horton-Sherman concept does not provide. This need gave rise to source area hydrology.

1. <u>The variable source area concept</u> (Hewlett and Hibbert 1967) was proposed by forest hydrologists about 1960 to account for the fact that neither stormflow nor baseflow is uniformly produced from the entire surface or subsurface area of a basin, as visualized in the Horton-Sherman concept. Instead the flow of water in a stream at any given moment is under the influence of a dynamic, expanding or shrinking source area, normally representing only a few percent of the total basin area. The source area is highly variable during stormflow. In the simple case of a permeable basin with a dendritic drainage network the source area pattern will appear as in Fig 7-12.

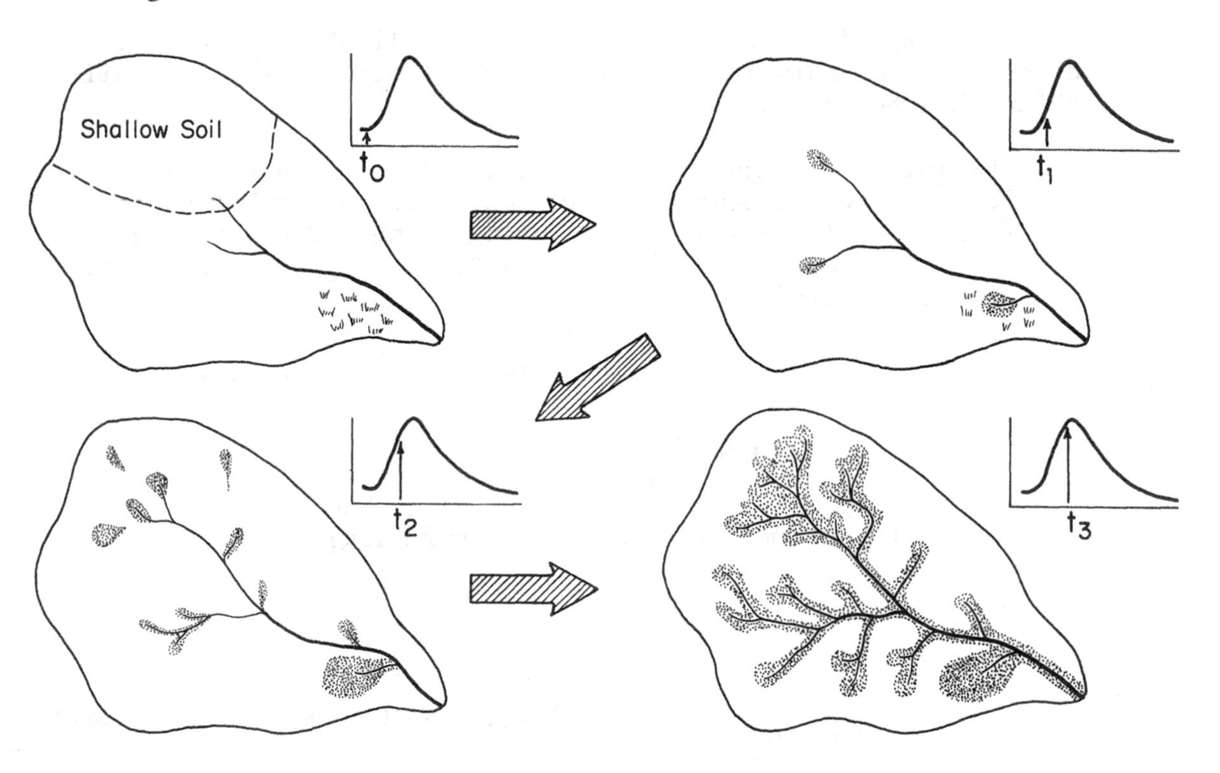

Fig 7-12. The small arrows in the hydrographs show how streamflow increases as the variable source extends into swamps, shallow soils and ephemeral channels. The process reverses as streamflow declines.

2. <u>Stormflow sources</u>. If the soil mantle of a drainage basin is deep enough to support vegetal cover through normal rainless periods, nearly all rainfall and snowmelt will infiltrate the soil surface before the water reaches a perennial, intermittent or

ephemeral stream. Whether the water remains in the soil to feed baseflow and evapotranspiration or re-emerges quickly to become stormflow, depends upon the depth and porosity (the dynamic storage capacity) of the soil mantle. Storage capacity is limited where the water content of the soil is approaching saturation and where layers of restricted conductivity exist close to the soil surface. The stream surface, channel banks and surrounding seep areas, and zones of restricted conductivity (for example, plow pans under fields adjoining streams) are the main source of stormflows. Except in regions of critically shallow soil mantles (Fig 7-2, central Kentucky), only extreme rainstorms exceed the dynamic storage capacity of forested watersheds.

The area actively involved in producing stormflow will vary from less than 1% of the basin during small storms to more than 50% in extreme storms (Fig 7-12). Cultivated fields, heavily grazed pastures and roadways become a part of the variable source area for stormflow whenever they discharge directly to stream channels. Stormflow and its source area increase at the beginning and decrease at the end of a rainstorm as a result of two simultaneous and virtually inseparable processes: subsurface flow, and channel expansion and shrinkage.

Subsurface stormflow. Precipitation into the perennial channel is the first water to cause a rise in the hydrograph. If rain continues, channel precipitation is augmented within a few minutes through expansion of subsurface areas that contribute water by (1) displacement of water stored in the channel banks and seep areas and (2) direct flow of new rain water through the large pores of the expanding channel banks. Overland flow that develops from relatively impervious rock, road or soil surfaces adjacent to the stream may be regarded as channel expansion. The solid black arrows within the soil mantle in Fig 7-13 represent the relative effect of a large rainstorm on stormflow as the distance upslope increases. Rainfall on the ridgetop contributes little or nothing to stormflow, although percolation within the soil profile (shown by the blown-up soil mantle sections) begins to displace stored water downslope. The latter water will either supply evapotranspiration or will add to baseflow in the weeks and months ahead. As in the case of baseflow, the lower slope continually receives soil water from upslope; the soil near the stream will always be wetter than upslope soil at the beginning of a storm.

Channel expansion. As rainfall continues to soak the slopes, the capacity of the soil mantle near the stream to transmit subsurface flow is exceeded and the water emerges at the surface further upslope and upstream. The emerging water quickly collects in draws to form intermittent and ephemeral channels which in turn increase the channel length during a large storm to perhaps ten or twenty times the perennial length. Channel expansion reaches quickly into

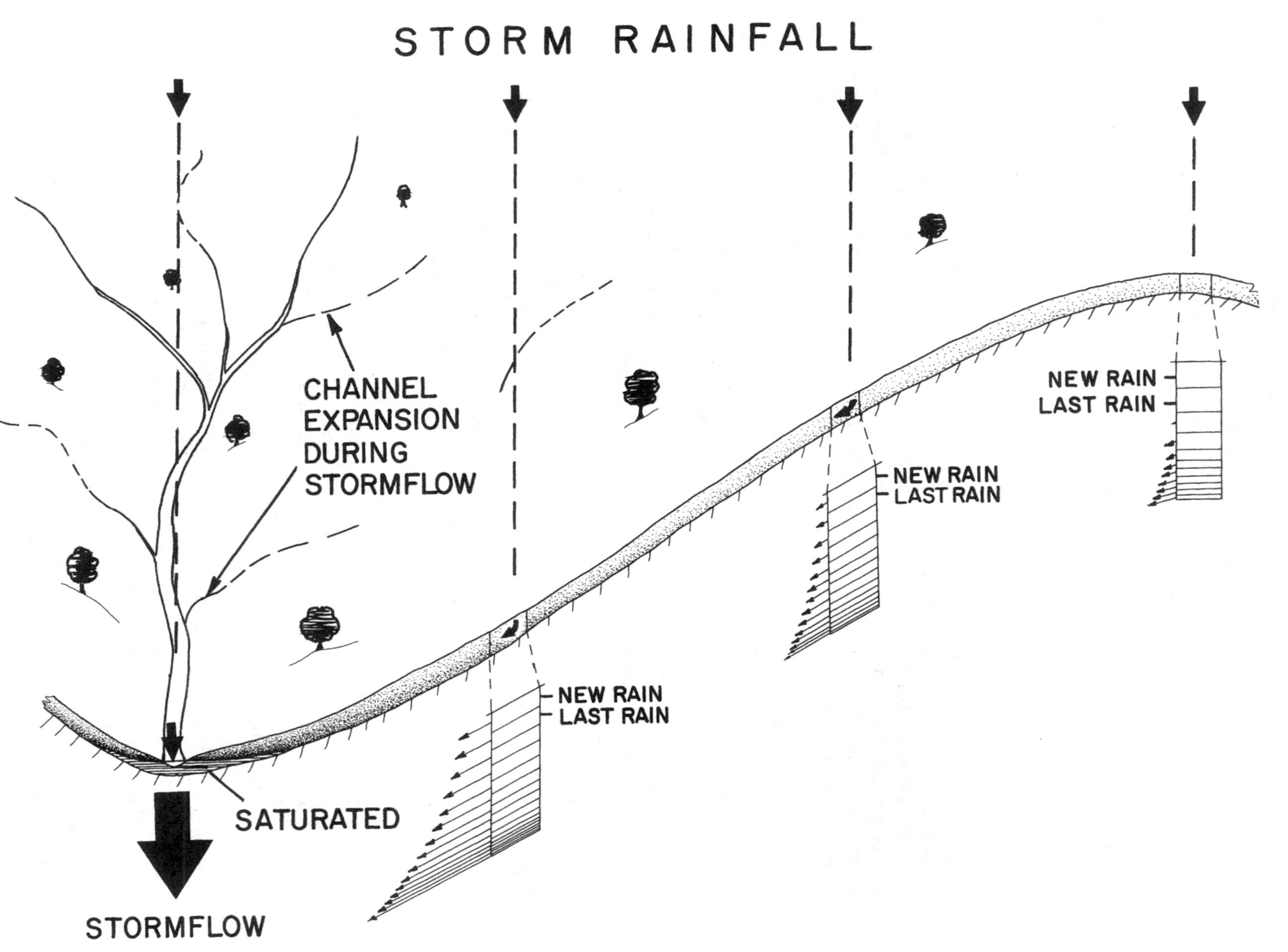

Fig 7-13. An idealized section through an upland forest basin, showing the varying source area for stormflow (direct runoff) and the source of delayed baseflow (from Hewlett and Hibbert 1967).

areas with shallow soils and bare rock surfaces, or into areas where overland flow occurs due to roads, trails and compacted soil. Exceptionally rapid expansion of the channel system gives the appearance of universal overland flow, but such a phenomenon is rare outside cities and cultivated fields. Reduced land slope does not change these relations in kind but only in degree.

The care of the channel system and its headwaters is a major responsibility of the land manager. Management prescriptions such as streamside management zones, devised to control non-point source pollution, are examples of management adjusted to variable source area patterns.

3. Baseflow sources. Under humid climates, any part of the soil mantle capable of storing soil water will provide some baseflow over a long period of time. The infiltrated water supply is normally greater than evapotranspiration; therefore, dynamic baseflow sources expand in wet periods and shrink during droughts. In drier climates, the limited rainfall is stored along the water divides until evaporated, leaving little to move toward the stream channel. Baseflow decreases rapidly in semi-arid lands because source areas are often limited to the intermittent channel and its banks.

In steep terrain, narrow ground water aquifers along the stream channel serve both as a limited storage body and as a conduit through which water is fed to baseflow. Much of the water passing through the narrow saturated zone comes from storage in the zone of aeration, where the water has been held at potentials between -5 and -200 cm. Although the drainage of unsaturated soil is relatively slow, large volumes of unsaturated soil below the immediate reach of evapotranspiration will yield water for months, even years, without further recharge.

In flat country with extensive ground water tables close to the soil surface, most baseflow comes from the shallower ground water, which responds to rainfall by emerging at the surface all across the basin. In coastal flatwoods, the zone of aeration is so thin that it serves as a minor source of baseflow.

In both terrains a greater percentage of the rain falling near streams will eventually become baseflow than of the rain falling near the water divides. Water infiltrated near the divide will reside longer in the soil mantle and will be subject to evapotranspiration on its way to the channel. The quality of the effluent baseflow will be related to residence time in the mantle and to the concentrating effect of evaporation. A mixture of old and new water will determine the quality of baseflow as it emerges from channel banks and springs. In general, old soil water dominates quality of baseflow.

As a hypothetical example, the response of nutrient export in baseflow to heavy doses of fertilizer at two points, A and B, on a basin slope may be represented:

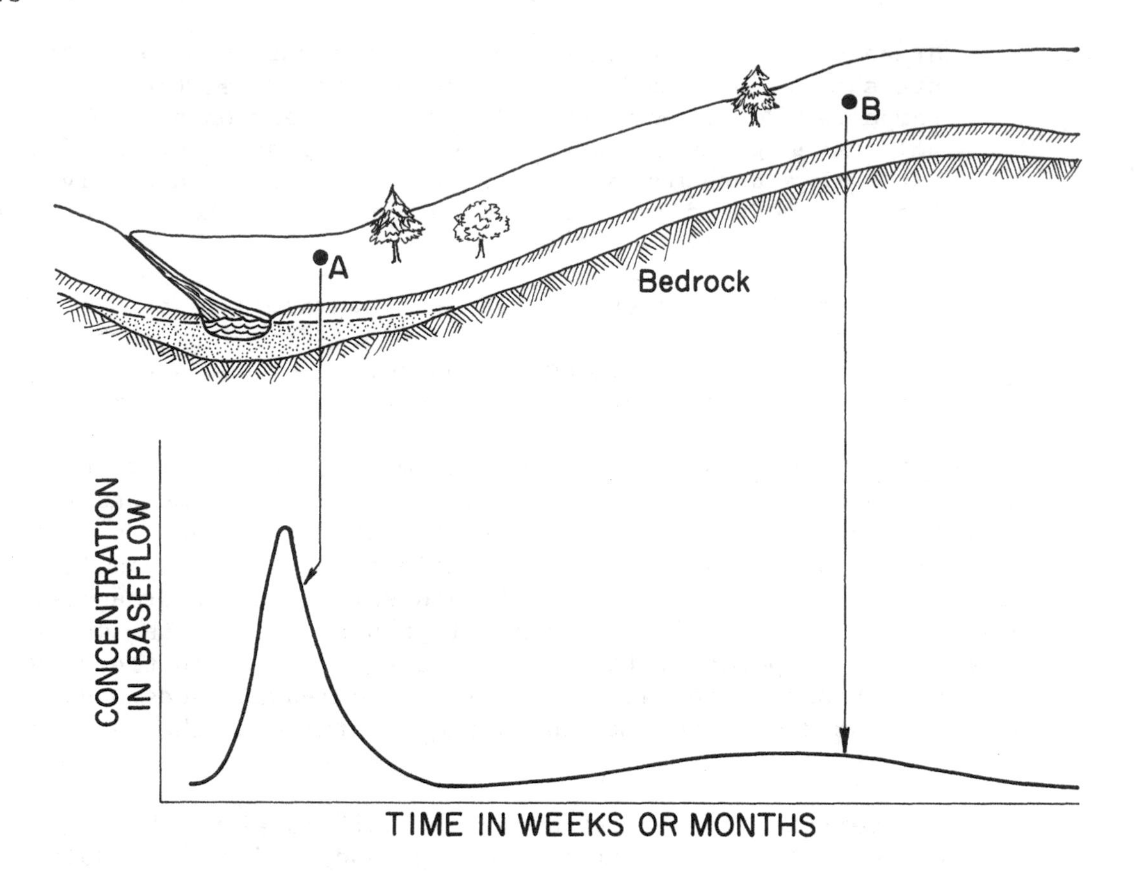

Remote placement of fertilizer at B will cause attenuation of its export or cause it to be retained in the basin. Mineralization of soil water near the water divides and the intervening evapotranspiration tends to concentrate minerals and nutrients downslope, where the additional water and nutrient will enrich the site. But minerals or pollutants added to soil along the stream may be quickly flushed out by storm and baseflows. Streamflow, minerals, nutrients, sediments and other pollutants are derived from variable source areas.

4. Computer models of the runoff process have been devised to simulate annual hydrographs and to estimate the frequency and magnitude of high and low flows in rivers with little or no hydrologic record. Hydrologic models may be classified many ways but one distinction important to forest and wildland managers is between those models that are infiltration-based and those that are based on the variable source area concept. Traditional engineering models are infiltration-based.

Traditional computer models use rainfall or snowmelt and a number of basin variables to simulate overland flow and route it to the stream reach of interest. Most operate as shown on the left in Fig 7-14. A small amount of subsurface stormflow (often called interflow) is allowed but model restrictions on channel expansion force most stormflow into overland pathways. The soil mantle is arbitrarily divided into an upper and lower zone to allow partly for dynamic subsurface storage, but current engineering models do not account for the variable source area process. Even though overland flow is greatly over-estimated, river hydrographs are rapidly and conveniently simulated for designing structures, forecasting floods and low flows, and operating complex reservoir systems. However the land manager should beware of interpreting source area processes (nutrient cycling, erosion, non-point pollution, and so forth) from engineering models.

Variable source area models by contrast seek to represent the actual pathways, residence times and sources of water as it flows through the headwater drainage basin. The flow chart on the right side of Fig 7-14 shows that these models are concerned with subsurface water movement (rather than infiltration rates) and with actual moisture contents by depth and position on the slope. Most stormflow comes from the expanding and shrinking zone of saturation at the base of slopes. Rainfall is assumed to infiltrate unless the surface layers become saturated, thus eliminating the virtually impossible task of estimating "excess rainfall" by theoretical infiltration models. Prediction of soil moisture content and its slope distribution become far more important than infiltration estimates in using a variable source area simulator. Detailed information about soil mantles and soil physical characteristics is also needed. Source area simulators will be useful in teaching runoff theory and in predicting the effects of land use on water quality, timing and yield, but detailed data requirements may restrict routine use in engineering applications.

Environmental regulations that impinge on land management are partly based on models of natural organic, mineral and water cycles. The land manager therefore needs a working understanding of these models and of the assumptions on which they are based. If the question at hand is the source or pathway of a pollutant (for example, an aerially-sprayed pesticide), interrogation of an infiltration-based simulator may suggest to the user that the source was the whole basin and that the major pathway was overland flow. Applied to the same case, a variable source area simulator would suggest that the source was the channel area and that overland flow was a minor pathway.

A case in point is the design of streamside protection areas in forest practice. When it is assumed that overland flow normally pours over the hillslopes during stormflow, "filter strips" of fixed width on either side of the perennial stream will be recommended. On the other hand, when it becomes recognized that the normal source of stormflow is the expanding network of intermittent and ephemeral channels, narrower and longer buffer strips will be designed to accommodate the dynamics of the varying source area.

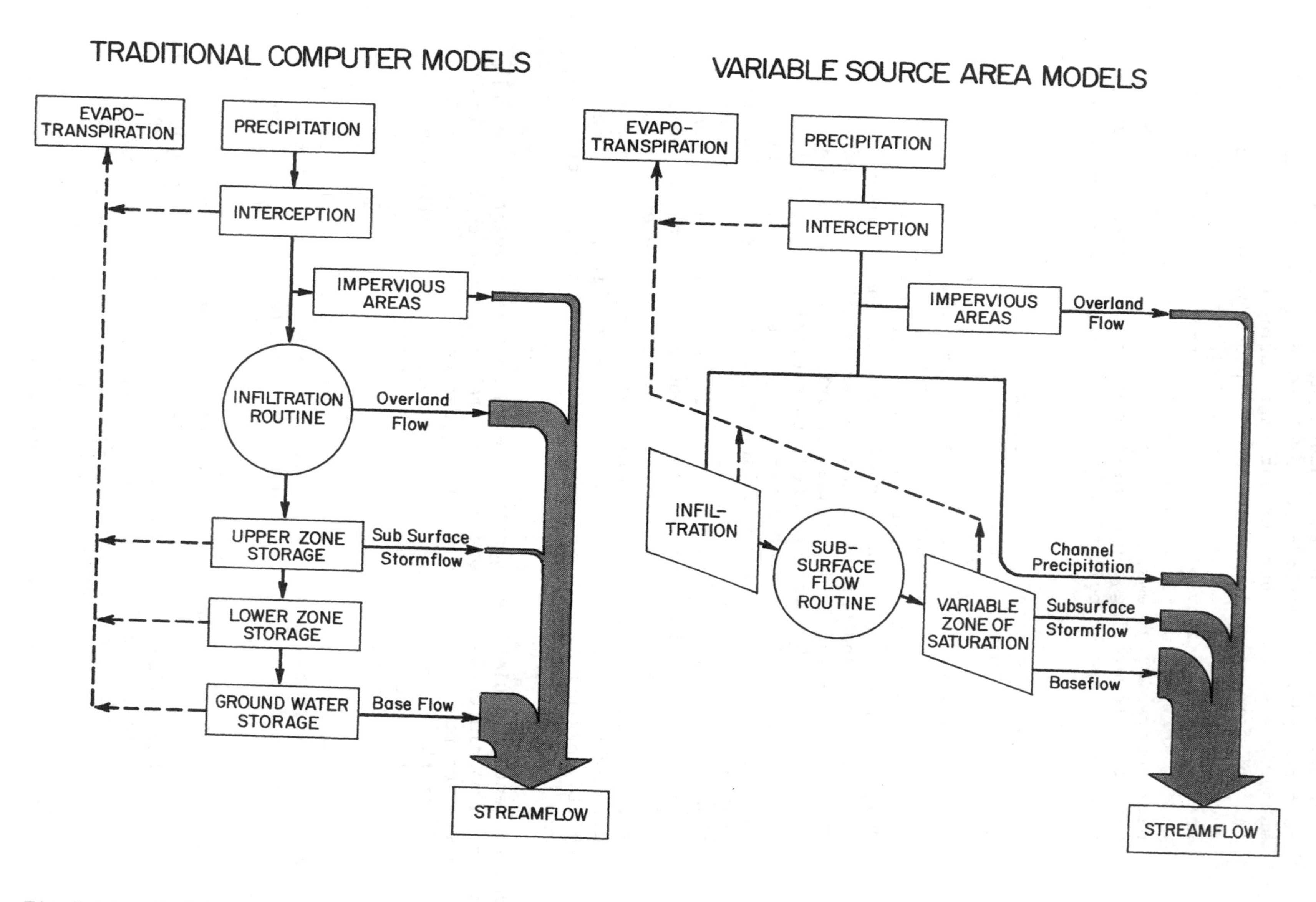

Fig 7-14. Models of the runoff process reflect varying views of how the source areas generate streamflow. The main difference between traditional and variable source area models lies in the way infiltration is dealt with. (Diagram by Pierre Y. Bernier)

Further readings.

Amer. Soc. Civ. Eng. Nomenclature for Hydraulics. ASCE-Manuals and Reports on Engineering Practice, No. 43, 501 pp., 1962.

Chow, V. T. Handbook of Applied Hydrology. McGraw-Hill Book, Co., New York, N.Y., Sections 14 through 25, 1964.

Hewlett, J. D., G. B. Cunningham and C. A. Troendle. Predicting stormflow and peakflow from small basins in humid areas by the R-index method. Water Resour. Bull., Vol. 13(2), pp. 231-253, 1977.

Hewlett, J. D. and A. R. Hibbert. Factors affecting the response of small watersheds to precipitation in humid areas. In: Intern. Symp. on Forest Hydrology, Sopper and Lull (Eds.), Pergamon Press, Oxford, pp. 275-290, 1967.

King, H. W. and E. F. Brater. Handbook of Hydraulics. 5th Ed. McGraw-Hill Book Co., N.Y., 32 pp., 1963.

U.S.D.A. Field Manual for Research in Agricultural Hydrology. Agriculture Handbook No. 224, U.S. Gov't. Print. Ofc., 547 pp., 1979.

U.S. Forest Service. An Approach to Water Resources Evaluation of Non-point Silvicultural Sources. U.S.D.A. and Environ. Prot. Agency, Wash., D.C., through Nat. Tech. Info. Serv., Springfield, Va., 760 pp., 1980 (Chap III).

S.C.S. SCS National Engineering Handbook: Hydrology. Sec. 4 Rev., U.S.D.A., Soil Conserv. Serv., Gov't. Print. Ofc., Wash., D.C., 547 pp., 1972.

Ward, R. C. Principles of Hydrology. 2nd Ed. McGraw-Hill Book Co., N.Y., 367 pp., 1975.

Problems.

1. Depth and velocity measurements at a stream cross-section will be handed out. Calculate the discharge Q in m^3/min and in $m^3/min/km^2$. The drainage basin area is ________.

2. Stage records and a discharge rating equation for a stream gaging station (weir) will be handed out. Compute and plot the hydrograph (m^3/min) for a stormflow period of about 2 days. Compute total discharge in m^3 for the same period.

3. The storm rainfall that produced the hydrograph in Prob 2 was ________ cm and the drainage basin area was ________ ha. Use a flow "separation" rule of 0.033 $m^3/min/km^2/hr$, compute graphically the volume of stormflow. What was the response R? Compare with Fig 7-2; what kind of basin might it be?

4. Given the information in this diagram and a Manning's n value of 0.05, compute the discharge Q and the mean velocity $\bar{V}$ using Manning's equation.

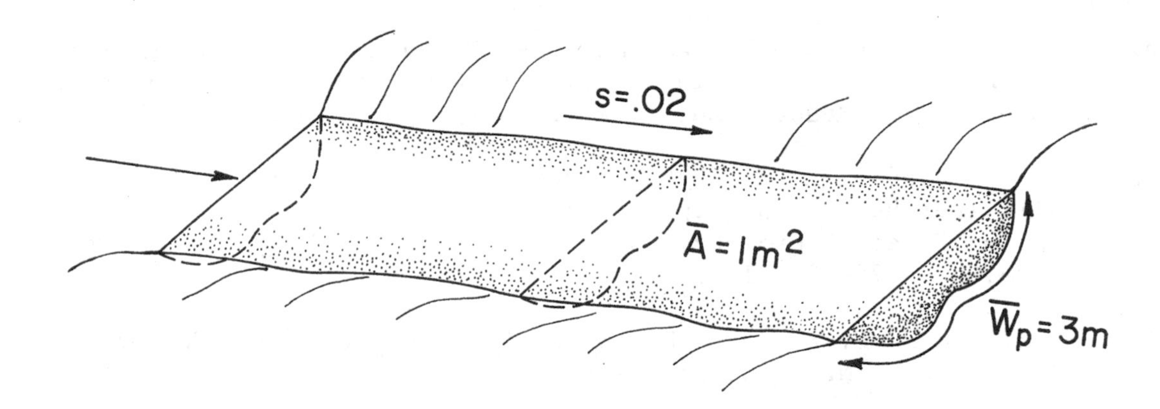

5. Prove that changing the coefficient C in Eq 7-8 from 1.34 to 2.48 converts Q from m^3/sec to ft^3/sec.

6. Use the unit graph in Fig 7-9 to predict the 100-ha basin hydrograph for these hourly excess rainfalls:

Hour	=	1	2	3	4	5	6	7	8	9	10
Excess rain (cm)	=	1.0	3.5	1.9	0	0	4.0	5.1	2.0	0	0.4

7. Route the outflow hydrograph in Fig 7-11 through another pond just below and identical to the pond in Fig 7-10. Plot the result in $m^3 \times 10^2$ per hour.

8. A 1-ha fish pond in the Upper Piedmont of the Southeast is shown in the topographic map on the next page. The average depth of the water below the overflow pipe is 1.5 m. Stormflow into the pond is cooler than the pond water and therefore tends to sink, pushing older water out by the overflow pipe. A fish biologist recommends that the pond be maintained at a given nutrient level during the summer by adding fertilizer. He asks what size rainstorm during June, July and August is just sufficient to displace all the water in the pond, thus reducing the nutrient level to

that of stormwater. Determine the drainage area above the dam in km^2 and use Eq 7-16 to give an approximate answer to his question. Assume an R from Fig 7-2 equal to 0.14.

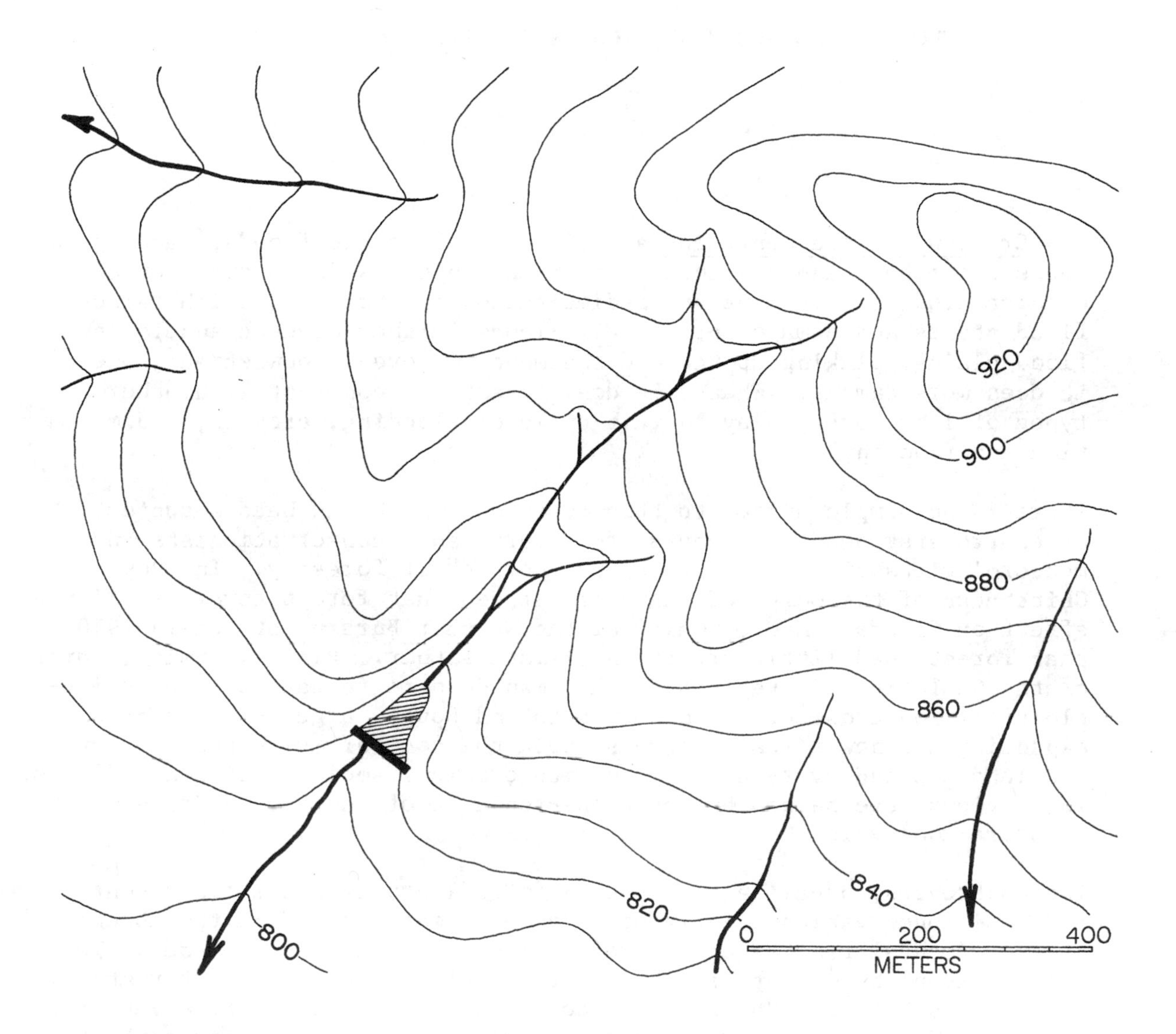

CHAPTER 8

EROSION AND SEDIMENTATION IN RELATION TO FORESTS

A. Erosion and sedimentation are closely related to floods; each aggravates the other, and all three are undesirable. Rains produce erosion, erosion produces sediment and sedimentation in channels, which raises flood stages and damages property. Floods further erode channels and flood plains, picking up the old sediment to move it downstream, where it does more damage. What role does forest, in contrast with other types of land cover, play in this cycle of flooding, erosion, sedimentation and flooding?

There is no simple answer to this question, which has been a subject of controversy among engineers, foresters, soil conservationists and meteorologists since the "propaganda period" of forestry. In 1909 Chittenden of the Corps of Engineers argued that forest cover had little effect on floods. Moore, Chief of the Weather Bureau, stated in 1910 that forests had little effect on climate either. Finally, Smith, Chief of the Geological Survey, said "What man does to forests will have little effect on erosion." Pinchot, Roth and Hough, arguing in favor of expanding the new forest reserves, took the contrary view that forests and land use had everything to do with erosion, sedimentation and floods. Their views were backed by the then-President of the United States, Theodore Roosevelt.

The controversy became entrenched in federal programs for flood control and soil conservation. Early disagreements were argued in the absence of experimental results and therefore stood little chance of being resolved except by weight of opinion or political power. The "forest enthusiasts," to use Chittenden's words, won their forest reserves in 1911 (The Weeks Law), and later an enlarged role in flood and soil erosion control in 1936 (The Flood Control Act). However, public confusion about the relation of forests to erosion and floods persists to this day. Part of the reason may be that resource managers and planners, as well as extension services, have at times relied on oversimplified explanations of erosion and floods in their efforts to guide the public. Examples of slogans with a plausible sound are: "Fire destroys watersheds"; "Stop the raindrop where it falls"; "Forests don't produce flood waters"; "Forest cover stops erosion." Some of these were meaningless jargon, others had an element of truth or a local application, but popular slogans tended to mislead the public and even to confuse land policy.

Terms must be defined before we discuss causes and effects.

1. Erosion is the process by which soil and minerals are detached and transported by water, wind, gravity, ice and man's activities. Physical energy, chiefly in the form of gravity or kinetic energy, and chemical energy, chiefly as a weathering process, underlie erosion in all its forms. This text is concerned mainly with water, gravity and man's activities as erosive agents. However, wind has reshaped land in desert and "dust-bowl" areas throughout time and ice is locally important in high mountains. Historically, ice and ice melt shaped much of the land, lakes and rivers north of continental glaciation.

Mass wastage refers to all forms of detachment and transportation by gravity alone. Included are land slips, slides, mud flows, rock falls, talus (rock spills at the base of cliffs), dry ravel (surface rocks rolling downslope following fires, soil thawing or drought periods), and soil creep (Brown 1980).

Soil creep is mass wastage by general downslope movement due to gravity stress on the weathered soil mantle. In geologic time, even solid rock is plastic and tends to deform under gravity and pressure. Under cycles of wetting and drying, hillslopes react similarly in a matter of decades, as the gradual leaning of large trees downslope indicates. In the headwaters of steep country, soil creep tends to develop convex-straight slopes (Chap 3); high water undercuts the bulge, gradually weakening slope stability and causing land slips into the stream channel. It is not fully known whether forest cover decreases this form of erosion by binding the creeping soil mass, or increases it by adding to the weight and thrust of the downslope gravity force, but there is ample evidence in northern California that soil creep and subsequent land slips occur on a massive scale on both undisturbed and cut-over forest land. Under humid climates, hill slopes in excess of 10 or 15% are very slowly moving their soil mantles downward, a process that ultimately leads to channel-related erosion even under forested conditions.

Surface water erosion is the detachment and transportation of soil materials by water striking or moving over the surface of the ground, or moving in perennial, intermittent and ephemeral channels, rills and gullies. The fact that soil has been detached and moved does not necessarily mean that the erosion products (sediments) have been exported from the land; export to the sea may require many cycles of detachment, transportation, sedimentation and detachment again, over a time span of centuries.

Subsurface water erosion is the elutriation of the soil mantle by subsurface waters, usually in the form of dissolved minerals. Colloids and silts may also be found in spring and seepage waters. It is not unusual for ground water to exceed 20 parts per million (ppm) dissolved matter. Unlike surface water erosion, the products of subsurface water erosion move quickly to the ocean. Some geomorphologists suggest that subsurface water erosion plays as large a role in shaping the land as surface erosion. Burrowing animals also perform subsurface erosion, not minor over geologic time.

2. Erosion hazard describes erosion potential by regions, localities and land use; it reflects the combined effects of erodibility and erosivity. Erosion hazard is naturally greater in temperate regions receiving only 20 to 50 cm of precipitation per year, a moisture supply sufficient to produce a number of erosive rainstorms each year but insufficient to maintain complete vegetal cover. An example is the short grass prairie region of western Kansas, Texas and New Mexico. To be useful in practice, the following terms require clear definition.

Erodibility is a characteristic of the material subject to erosion that denotes susceptibility to erosive agents. Sands are generally more erodible than silts, and silts than clays, but no fully satisfactory way to predict soil erodibility has been found. Changing conditions of the soil (soil wetness, soil frost, recent tillage or compaction) change the erodibility of soil. Angular soil particles are more interlocking than rounded particles; soil colloids cement particles together; compaction increases total surface contact among particles. All of these tend to reduce erodibility. Another important factor is whether water is infiltrating or exfiltrating the soil surface while detachment under water is taking place; the filtration force reduces erodibility in the first instance but increases it in the second. Gullies often begin along midslopes where water exfiltrates during rainfall.

Erosivity is a characteristic of the erosive agent, i.e., water, wind or ice. High velocities increase erosivity as well as the type and amount of suspended sediment. Falling rain is more erosive than water moving over the surface of the ground. Because of larger drop size, higher velocities and intensity, rainfall in the southeastern part of the United States is 6 to 10 times more erosive than rainfall in New England (Wischmeier and Smith 1965).

Natural erosion (geologic erosion) generally refers to erosion rates that are virtually unaffected by man and his activities. Vast expanses of sedimentary rocks and the deltas of some major rivers testify to massive erosion throughout geologic time.

Accelerated erosion refers to erosion rates that are man-caused or man-related. In general, accelerated erosion is in addition to natural erosion, but separation is often difficult.

Cultural erosion is the detachment and transportation of soil materials by mechanical devices used in agriculture and construction work, particularly plowing, road building, channel dredging, surface mining and land filling. In this case, man is the direct agent producing erosion. If improperly done, nearly all such activities lead to further erosion by water, gravity or wind.

Erosion pavements are layers of stones or pebbles on the soil surface which both indicate past erosion and serve as a barrier to further erosion. Common in semi-arid areas, erosion pavements are made up of coarse material formerly within the layer of soil that has eroded away. Estimates of natural erosion have been computed by relating the amount of stones in the erosion pavement to the average stone content of the underlying soil.

3. Products of erosion. After soil, minerals and debris are set in motion, different terms are used.

Suspended sediment is particulate matter suspended in and carried by moving water. On average, less than 3% of the total mass of streamflow is suspended sediment. The Yellow River of China is said to reach as high as 40% suspended sediment during heavy floods.

Sedimentation is the process by which materials carried in water are deposited. Materials are considerably mixed, sorted and segregated by size during the process; coarse particles move relatively short distances and finer particles move longer distances. Very fine clay in suspension will move to the nearest body of still water, where it may form a fine layer uniformly over the bottom.

Dissolved load refers to materials carried in solution by water, including organic and gaseous solutes. Dissolved load and its quality determine the biological productivity of streams.

Bed load refers to coarser materials, usually mineral, moving along the bottom of streams during both stormflow and baseflow. Gravel, stones and boulders are subject to forces of velocity and bouyancy which tend to move them along by skipping (called saltation), tumbling, rolling and sliding. In steep mountain streams, much of the total stream load during stormflow is bed load. In streams with sandy bottoms, baseflow also moves sand particles along by saltation. Bed load is the most difficult part of erosion to estimate.

Organic debris includes leaves, roots, twigs, bark, fruits and stems, as well as animal parts. Normally not thought to be part of the erosion process, such matter nevertheless directly affects erosion of soil from detachment to basin export and it becomes part of the total stream load. When deposited on flood plains, organic debris favors soil fertility and rapid vegetal growth, which in turn aids in the trapping of more sediment. Aquatic fauna and fungi decompose debris in streams and assimilate much of the nutrient.

Total stream load includes all the organic and inorganic materials carried past a designated point along a stream channel, either as solutes, suspended sediment, bed load or organic debris.

B. Mechanics of water erosion. Erosion proceeds in three steps: Detachment, transportation and deposition (sedimentation). These steps are integrated but it aids understanding to think of them separately.

1. Surface erosion is caused chiefly by water (wind erosion is beyond the scope of this text). Detachment occurs because of impact, breaking, plucking, bouyant lifting, freezing-thawing or chemical decomposition. Erosive agents are falling rain, flowing water, glacial ice, soil frost and the chemical activity of water. In simple terms, the goal of management is to minimize detachment by dispersing water energy.

Colloidal detachment, dispersal and solution begin the process, largely under the influence of raindrop energy, but aided by temperature fluctuations and soil freezing.

Splash erosion under raindrop energy transports soil particles both up and downhill. Prevailing wind during rain affects the net direction of soil movement in open country but, in general, land slope is a more important influence; at 10 percent slope, about three times as much soil splashes down as up.

Sheet erosion occurs by a combination of splash and surface water movement downhill. Much of the soil transportation formerly attributed to sheet erosion is actually due to rill erosion under repeated cultivation, which each time wipes out old rills and gives the false impression of sheet-like ablation of the land.

Rill erosion occurs when surface water moves into small depressions, gains depth and velocity, and begins the transportation of erosion products. A dendritic or trellis network of rills finally produces enough depth and velocity of flow to initiate gully formation. A sharp break in the slope gradient, or scattered debris from crops or forest, will cause a drop in velocity and deposition of some suspended sediment.

Gully erosion may proceed rapidly because of increased depth, volume and velocity of water. Gullies usually begin to form where subsurface stormflow emerges on the hillslope. As the gully cuts downward, the slope into it may become vertical. Two types of gully extension may be distinguished:

V-shaped gullies form in soils that are either shallow or uniformly erodible with increasing depth. Loessial (windblown) soils often exhibit this form of development.

The process continues with splash and rill erosion on the convex gully walls. The gully bottoms remain strongly sloped toward the outlet. Vegetal cover is usually sufficient to control V-gullies because raindrop energy is the prime erosive agent.

U-shaped gullies form in soil that has less erodible soil overlying more erodible soil, as is the case where a fine textured B horizon caps a sandy C horizon. Subsurface flow or water trickling down the gully walls softens the sandy subsoil, causing it to cave in. The less erodible soil above holds firm until well undercut, whereupon the entire wall collapses (mass wastage) into the gully channel. Subsequent rainfall transports the detached material easily along the almost flat gully bottom. U-shaped gullies frequently erode through the water divides above them, proving that very little surface water movement is required to keep the gully active. Structures and vegetation are usually required to control U-gullies (Heede 1976, for further information).

Channel erosion occurs by recutting old deposits, or by natural or man-caused meandering of streams. The flood plains of major streams are slowly reworked by the kinetic energy of water. Much of the energy of small streams is dissipated in bed load movement and abrasion of rock. The entire subject of fluvial geomorphology deals with channel erosion processes (Leopold et al. 1964).

Abrasion occurs in all channels due to the solid materials moving with or suspended in water. Abrasion in long rivers grinds up stones and boulders until chiefly fine sediment and solutes are delivered to the oceans. Abrasion also serves to wear down "nick points" (bed rock outcrops in stream channels) which otherwide would tend to permanently stop the geological base-leveling of channels.

2. Detachment and transportation. All the above processes of surface erosion are due to the transformation of potential energy (energy of position) into kinetic energy (energy of motion). The kinetic energy of a falling raindrop is sufficient to kick soil particles a meter into the air. The kinetic energy equation is:

$$K_e = 0.5 M V^2 \qquad \text{UNITS: } ML^2T^{-2} \qquad 8\text{-}1$$

K_e is the energy of motion, M is raindrop mass, and V is raindrop velocity. Soil particles cling to each other by strong forces of adhesion and cohesion. Falling rain works mainly to detach soil particles; the resulting sheet, rill, gully and channel flow works mainly to transport them.

Example: Assume that 2 cm of heavy rain falls in 1 hr on 1 m^2 of bare soil. Drop velocity V_r is 7.6 m/sec (Chap 4,

Sec 3). The water fails to infiltrate and gains a downslope velocity V_s of 0.2 m/sec as surface water flow. Since the mass of water is the same, it is easy to compute the ratio of rainfall energy (K_{er}) to surface water flow energy (K_{es}). Using Eq 8-1:

$$K_{er}/K_{es} = (.5\ M\ V_r^2)/(.5\ M\ V_s^2) = V_r^2/V_s^2$$

$$= (7.6\ m/sec)^2/(0.2\ m/sec)^2$$

$$= 1444$$

In this hypothetical case, rainfall delivers 1444 times the energy of an equal amount of surface water flow. If any water infiltrates the ratio would be even larger.

The example is oversimplified because variations in rainfall intensity and drop size are not accounted for. Experimentally Wischmeier and Smith (1965) found that the energy delivered by a rainstorm can be estimated hour by hour from rainfall intensity I in in/hr (all references to this work are in the ft-lb-sec system):

$$K_e' = 916 + 331 \log_{10} I \qquad 8\text{-}2$$

K_e' is in foot-tons per acre-inch of rain (ML^2T^{-2}). Computed this way, rainfall energy is still 520 times the surface flow energy of an equal mass of water.

Most of this energy is dispersed in compaction, turbulence and heat. The interception of rain by vegetation coalesces drops into larger drops which, when falling from tree crowns, regain terminal velocity in about 7 m. Therefore only lower vegetation, such as brush, kudzu, grass, crop residues or forest debris, serves effectively to protect mineral soil from erosive raindrops. Splash erosion under high forest canopy may be severe if the forest floor has been repeatedly burned, trampled or removed. In some areas, grazed or annually burned woodlands deliver surface water downslope (overland flow if it reaches the stream) because the entrainment of fine soil by the energy of throughfall prevents infiltration.

Approximately 1 metric ton of organic debris per hectare will absorb 98 percent of rainfall energy. Forest floors usually vary from 1 to 5 tons/ha. However, a few hundred kg of debris, particularly fibrous materials such as pine straw, spread over a hectare of bare soil will reduce rainfall erosivity 75% or more. Litter detention storage further disperses water energy that otherwise would break the bonds between colloids and fine aggregates to produce a thin slurry of mud over the surface. Where exposed directly to raindrop energies, infiltration capacity is quickly impaired as mud clogs surface pores. Some evidence of local surface water movement caused by shifting of fine materials during large rainstorms may be found under full forest cover, but the hydrological effect is usually minor.

3. Predicting erosion. Erosion hazard has been reduced to six basic factor complexes in a "Field Soil-loss Equation" proposed by W. H. Wischmeier about 1960 (summarized in Wischmeier and Smith 1965). The equation serves as a mnemonic device for the factors underlying erosion hazards in agriculture:

$$A = R\,K\,L\,S\,C\,P \qquad 8\text{-}3$$

A is estimated field soil loss in tons/acre/year, R is precipitation erosivity, K is a soil erodibility factor evaluated by soil type, L represents the length of the slope contributing erosion products, S represents field slope, C is a crop management factor (stubble mulching, etc.), and P is an erosion control factor (terraces, etc.). Only A, K and R have dimensions in the equation; the other variables are expressed as ratios of expected soil loss to experimentally measured loss on "unit plots."

Wischmeier writes, "A unit plot is 72.6 feet long, with a uniform slope of 9 percent, in continuous fallow, tilled up and down the slope." Soil loss from such plots has been measured for over 35 years at various field stations of the USDA Agricultural Research Service. Soil loss rates from unit plots per unit of R served as the measure of K (soil erodibility K = A/R) for various soils studied. R times K is a measure of erosion hazard uncorrected for field conditions: Given a certain rainfall erosivity and soil erodibility, the unit plot will yield RK tons/acre/year. As L, S, C and P are varied from a value of 1.0, the estimate of soil loss is modified up or down from the fallow condition and a local prediction in tons/acre/year is made. Application of the method involves a number of tables and procedures too detailed for treatment here (refer to Wischmeier and Smith 1965).

The erosivity factor R is the annual sum of the product of two rainfall factors measured in each locality. Many years of study suggested to Wischmeier that two variables correlated best with soil loss from unit plots:

$$R = \sum_{1}^{n} K'_e\, I_{30} \qquad 8\text{-}4$$

K′ is computed by Eq 8-2 for each rainstorm, I_{30} is the maximum 30-min intensity (in/hr) within each rainstorm, and n is the number of storms per year. Eq 8-4 reminds us that the kinetic energy of rain detaches soil, and the turbulent surface flow produced by an excess of rainfall over infiltration moves soil from the field. A map of Wischmeier's R for eastern United States is shown in Fig 8-1.

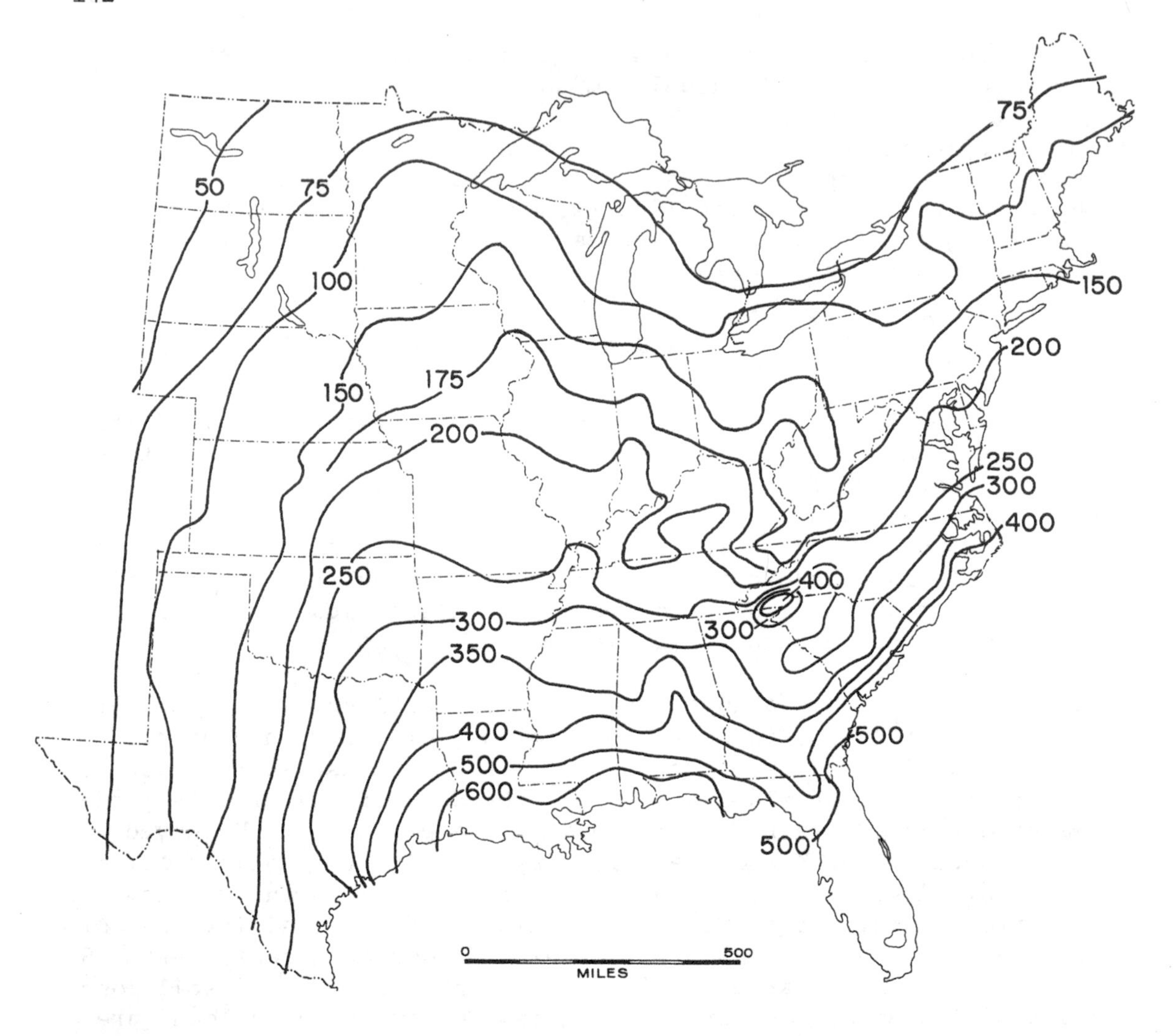

Fig 8-1. Annual values of Wischmeier's R (English units) for eastern U.S., called an "iso-erodent" map (Wischmeier and Smith 1965).

Forest and wildlands. While satisfactory to show the basic nature of erosion and to predict field loss under agriculture, Eq 8-3 needs modification for predicting soil loss from whole drainage basins or forest areas. The factors L, S, C and P are defined in terms of uniform slopes, crop spacing and soil conservation works. The forest briefly resembles the field situation only when root-raking or disking are used in site preparation. In most forest practices, the irregular pattern of soil disturbance and the frequent interruption of surface flow paths by debris will render sediment predictions by Eq 8-3 too high by an order of magnitude if applied over first-order drainage basins. Predictions of soil loss from dirt roads and decking areas serve as soil detachment indices but not as sediment export estimates. There is an urgent need for erosion equations to accommodate entire drainage basins and the complex pattern of forest and wildland practices.

Research may soon provide modified versions of field soil loss equations for use in forest practice. Modification may take this form:

$$A = R\,K\,S\,W$$

Where A, R and K are the same as in Eq 8-3, S is a modified slope factor and W is a forest operational pattern factor that will reflect the probability of a detached soil particle being exported during the year within which it was detached. W varies from near zero when the exposed soil patch is near the water divide, to near 1 when the exposed soil patch is at the stream channel. Obviously the probability of export is high if soil disturbance is concentrated near streams and low if concentrated near water divides. Management according to drainage basin pattern is the key to sediment control in harvesting, site preparation and planting operations. State environmental protection divisions now recommend some form of zoning in guides to "best management practices" for forestry (for example, Erosion control on forest land in Georgia, Hewlett et al. 1979).

4. Subsurface erosion is caused mostly by water; however, burrowing animals (including moles and termites) contribute to the process. Dissolved minerals in effluent water accounts for most subsurface export. Operating over millions of years, the slow leaching of porous soil and rock materials has shrunk the land surface and salted the sea. Because of the weak acids formed in rain and percolating waters, the process of subsurface erosion proceeds rapidly in limestone terranes forming sink holes and underground streams. The process is much slower in granites, basalt, shales and sandstones. The rate of subsurface erosion is subject to little managerial control, except to the extent that vegetal cover and land use alter the excess of precipitation over evapotranspiration; if more water is induced to percolate through soil mantles by reducing evaporation, then more energy will be applied to elutriation.

Soil piping is a form of subsurface erosion in which holes from a few cm to a meter in diameter develop in stream banks, arroyos and gullies. The process is associated with temporary saturation, high sodium content in soils, and in some cases is initiated by animal burrows. In areas of the Southwestern United States, soil piping is a major cause of gully expansion (Heede 1976).

C. Measurement of Erosion. Erosion is among the most difficult hydrologic factors to measure. Rainstorms and flood flows are so variable that one unusual event can deliver up to 75% of all the erosion products yielded by a small basin in a 10-year period. Large rivers will deliver sediment more uniformly in time but because much sediment moves as bed load, measurement of rates is still difficult. Measurement of erosion in source areas can be related to the cause, but measurement at the mouth of larger streams only establishes the general level of recent erosion in the watershed. Because of the urgent managerial need to separate accelerated from natural erosion, much research effort has gone into methods of measurement, which may be classified according to whether they are applied at the site of detachment, transportation or deposition.

1. Site of detachment. The movement of material from the source area may usually be related to the cultural practice or natural condition that causes the problem. Some methods are:

Iron pins serve as bench marks for periodic measurement of denudation relative to the original surface. Periodic photographs are taken from fixed points, annually or across decades. These are useful in studying gully development.

Rods driven deeply into sloping soils measure soil creep across years and decades; the downward leaning or deformation of the rod reveals the rate.

Tracers (dyes, radioactive materials and white sand) have been used occasionally to determine the exact pattern of detachment and removal of surface soils and stream bottoms.

The weight of an erosion pavement per unit area divided by the weight of stones per unit volume of subsoil is used as an index of geological erosion.

Soil profile studies, e.g., estimating the percentage of the A horizon remaining after cultural history, are sometimes used on a regional basis.

2. Materials in transport. Suspended sediment and bedload movement are estimated by taking samples of flowing water, or by trapping the material in a settling pool, or by a combination. With the addition of forest debris to the total stream load, this becomes surprisingly difficult. Samples must be taken proportionately to discharge rate. Bed load and organic debris can be estimated effectively only by stopping all the material and sub-sampling for analysis of organic and inorganic content. One such installation on a 100-ha basin is shown on next page.

Depth-integrated samples of suspended sediment must be used in large streams to assure that the entire flow past the point is represented. Particle size and sediment concentration will increase from top to bottom of the stream. The sample is integrated over depth by lowering and raising the sampler at a uniform rate so as to fill the sample bottle just as it emerges from the water surface.

Grab samples are taken quickly at a turbulent point in the stream where mixing of the water and suspended sediment can be assured. The grab sample is acceptable only on small

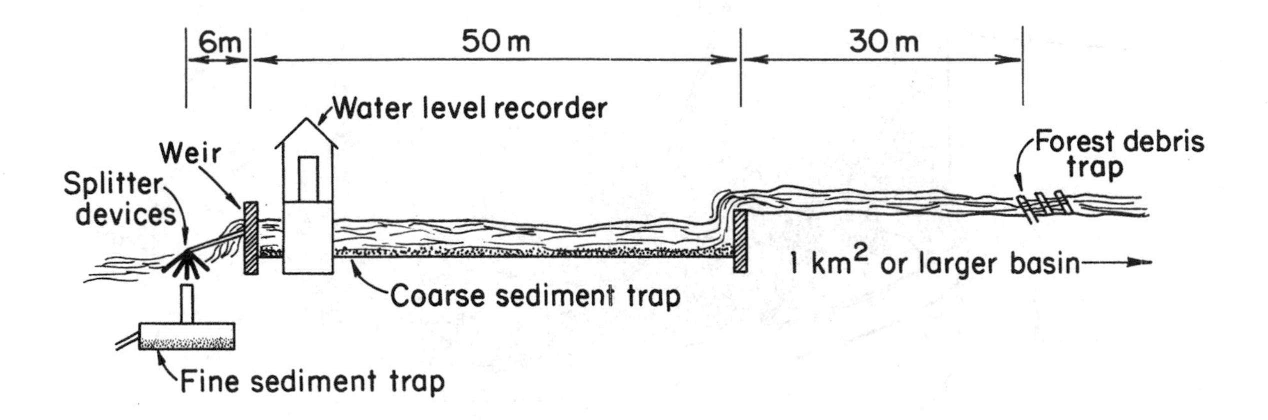

streams, or where dissolved materials are the only objective of measurement. The grab sample is representative of dissolved materials because differential gravity forces on solutes and water are negligible.

Bed load traps in stream bottoms are unreliable in large flood flows because of overfilling and violent currents around the trap. During baseflow and small stormflows, traps can be useful indicators of normal bed load movement. However, they must be frequently measured and emptied.

Automatic devices for sampling solutes and suspended sediment are now widely used for monitoring pollution and natural water quality. Some are complex, involving electrically operated pumps and servo devices; others are self-operating. One commonly-used proportional sediment sampler is the "Coshocton wheel" which is powered by the flowing water and diverts five-thousandths of the controlled flow into a storage tank for analysis (diagram on next page).

Turbidity of water is often measured to estimate its quality for recreational purposes, to monitor land use effects or to predict the penetration of sunlight to lower depths, the latter important in limnological work. One measurement consists of the depth of water through which a white disk can be seen in daylight; the disk is lowered on a rod until it just disappears. Turbidity, therefore, is defined as the attenuation of light through water caused, in this case, by suspended sediment. Turbidity is also measured by electrophotometers and Jackson Turbidity Meters, the latter using a standard candle as a source of light for the test. Turbidity is an index only; it does not measure total stream load.

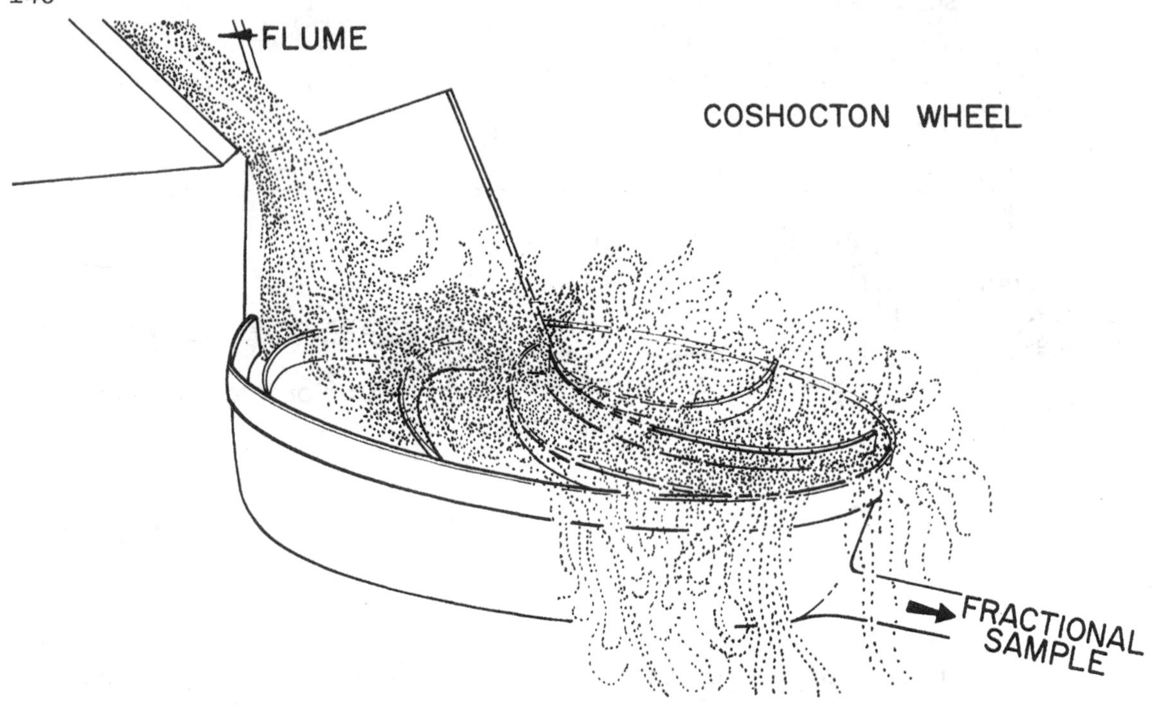

3. Mechanics of transportation. The energy of flowing water is less than that of falling rain, but is great enough to move erodible or already detached soil. Hydraulic studies show that the average velocity ($\bar{V}$) of water flowing in a rill, gully or stream will increase with the square of the depth (d):

$$\bar{V} = f(d^2)$$

The ability of water to move loose particles depends on their size (diameter of the particle). Eq 8-1 states that energy delivered to an exposed particle of a given size varies directly with velocity squared:

$$\text{Energy} = f(\bar{V}^2) \qquad 8\text{-}5$$

Manning's formula (Eq 7-3) states the velocity of water varies directly with the square root of channel slope (s):

$$\bar{V} = f(s^{1/2}) \qquad 8\text{-}6$$

Substituting in Eq 8-5, we see that water energy is a function of slope:

$$\text{Energy} = f(s) \qquad 8\text{-}7$$

Neglecting turbulence (an energy dissipator), the size of particles moved increases 4 times as water velocity doubles, but only 2 times as slope doubles. In other words, slope is not as powerful an indicator of erodibility as depth and velocity of flowing water. This fundamental principle is frequently overlooked and erosion formulae are often made too slope-dependent. Land slope is far less critical in forest practice than in field practice because forest debris reduces mean water velocities and encourages infiltration, which in turn reduces depth of flow. Erosion hazard indices in forestry should be based less on slope and more on pattern of disturbance as it affects flow paths and soil exposure.

4. Sediment deposition. The best way to develop regional denudation rates is by periodic survey of deposited materials in ponds, lakes and reservoirs. Reduction of water velocity in large bodies of water causes all but the solutes and colloids to settle out in deltas. The initial topography of the bottom is surveyed and subsequent sediment delivery is measured by rods or sonar soundings from boats.

D. Continental sediment export. The annual export of sediments to the oceans by the continents was reported by Holeman (1968):

Continent	Metric tons/km^2/yr
Africa	25
Europe	32
Australia	40
South America	56
North America	86
Asia	536
Weighted world average	182

Only Asia (China and India) exceed North America in average annual denudation. The U.S. contributes most of the N.A. average. Sediment yield of major U.S. rivers was summarized by Holeman (1968):

River	Metric tons/km^2/yr
Mississippi	97
Missouri	159
St. Lawrence	3
Colorado (before dams)	380
Columbia	35
Ohio	69
Brazos (Texas)	350
Alabama	34
Potomac	60
Delaware	51
Eel (Scotia, California)	2000--3000

Because of reservoirs, farm abandonment and improved soil conservation practices, the U.S. sediment yield has decreased since 1970 (Fig 8-2). The largest sources of sediment are row-crop agriculture, overgrazing and construction work. Forest practices supply only a small fraction of the total but sediment export from forest access roads and skid trails is locally severe.

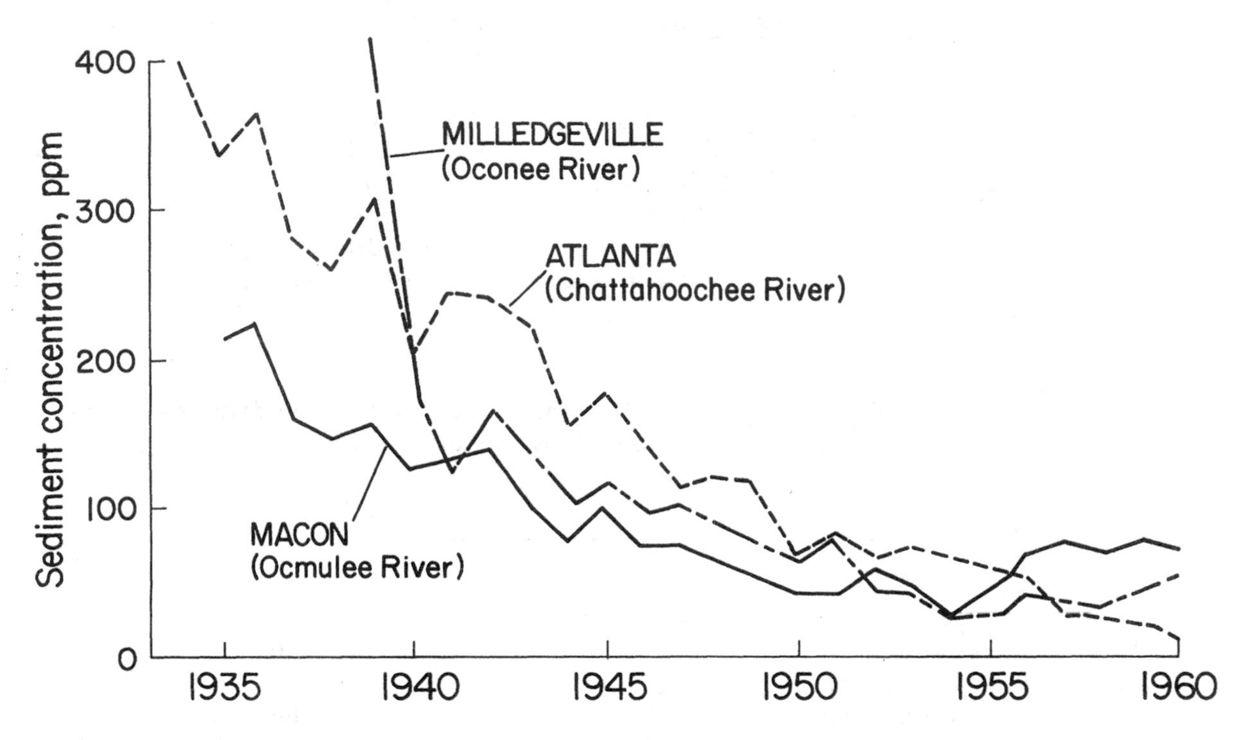

Fig 8-2. Sediment trends in Georgia (Yearbook of Agriculture 1955).

Sediment deposition destroys property, raises flood levels by reducing channel capacities, interferes with aquatic life and silts up waterways, reservoirs and natural lakes. Added to damages at the point of detachment (soil sterility), the total cost of natural and accelerated erosion is in billions of dollars annually.

How much erosion man can afford to stop is a top-level policy problem. Farmers talk of 5 tons/ac/yr (11.2 metric tons/ha/yr) of soil loss from fields as "tolerable." Foresters talk of 1 ton/ac/yr (2.24 metric tons/ha/yr) as tolerable. These tolerance limits exceed the world average rate of continental denudation quoted above. The soil erosion problem is not yet clearly identified: Its identification and solution is a prime issue in land policy.

E. Detachment rates are usually greater than sediment rates, depending on where the export is measured. Detached soil may travel a few meters from a skid trail or a field, settle out and remain in place for hundreds of years before moving again. Holeman (1968) reports that the annual export of sediment by the Potomac River (60 metric tons/km^2) is only 5% of the soil detached (eroded) within the basin. The smaller the area considered, the higher the ratio of export to detachment.

Export-detachment ratios are not highly sensitive to forest land use. For example, undisturbed and clearcut forest land may have the same ratio because both detachment and export are small in undisturbed forest while both are larger under practices that bare the soil to rainfall.

The managerial difference between detachment and export is that detachment is associated with on-site damages, while export is associated with off-site damages. On-site costs are borne by the land owner but off-site costs are borne by the neighbors or the public. Off-site costs have stimulated public laws to regulate on-site management, and the land manager is now compelled both by self and public interest to learn the exact causes of detachment and export of sediments.

1. Accelerated erosion. The great delta of the Mississippi River shows that natural erosion was immense before the arrival of Europeans in the tributary basins. The evidence is conclusive that accelerated erosion since then has been due, not to forest cutting, but primarily to 1) unwise cultivation and 2) excessive grazing. Today, construction work is adding substantially to accelerated erosion, including surface water erosion, cultural erosion and mass wastage. Some cultural erosion (movement of soil by machinery) is temporary; cuts and fills may be quickly stabilized by structural and vegetal means. Other earth-moving activities, such as levees, bottomland fills, channel dredging and mountain-topping, lead to later erosion by mass wastage and channel cutting.

Forest activities that accelerate erosion may be listed in order by problems to forest managers:

Road and skid trail layout, construction, use and maintenance affect erosion from forest land more than all other forest activities combined.

Channel encroachment by roads, harvesting, site preparation and tree planting is usually associated with poor access. Up to 90% of all the sediment exported from forest operations is due to poor roading and channel-zone practices (Fig 8-3). Camp sites and heavy fishing pressure are locally damaging to channels and deliver sediments directly to streams.

Site preparation methods, particularly root-raking and disking, are usually secondary as a source of sediment export. Tree planting by machine, when done on slopes, can rival site preparation as a source of sediment.

Harvesting activities, including felling, skidding, decking, loading and hauling, contribute sediment mainly when associated with bad access systems.

Fire prevention and suppression activities, particularly plowed fire breaks, are locally serious.

Recreation activities, such as camping, hiking, off-road vehicles, ski-lifts, summer home access and supporting activities, are causing increasing erosion and site damage, but the quantities of sediment exported are small.

Flatwoods drainage has caused local sediment problems.

Wildlife management activities, including overpopulation by ungulates, stream "improvements," careless access by hunters and fishermen, and hunt camps, are locally serious sources of sediment.

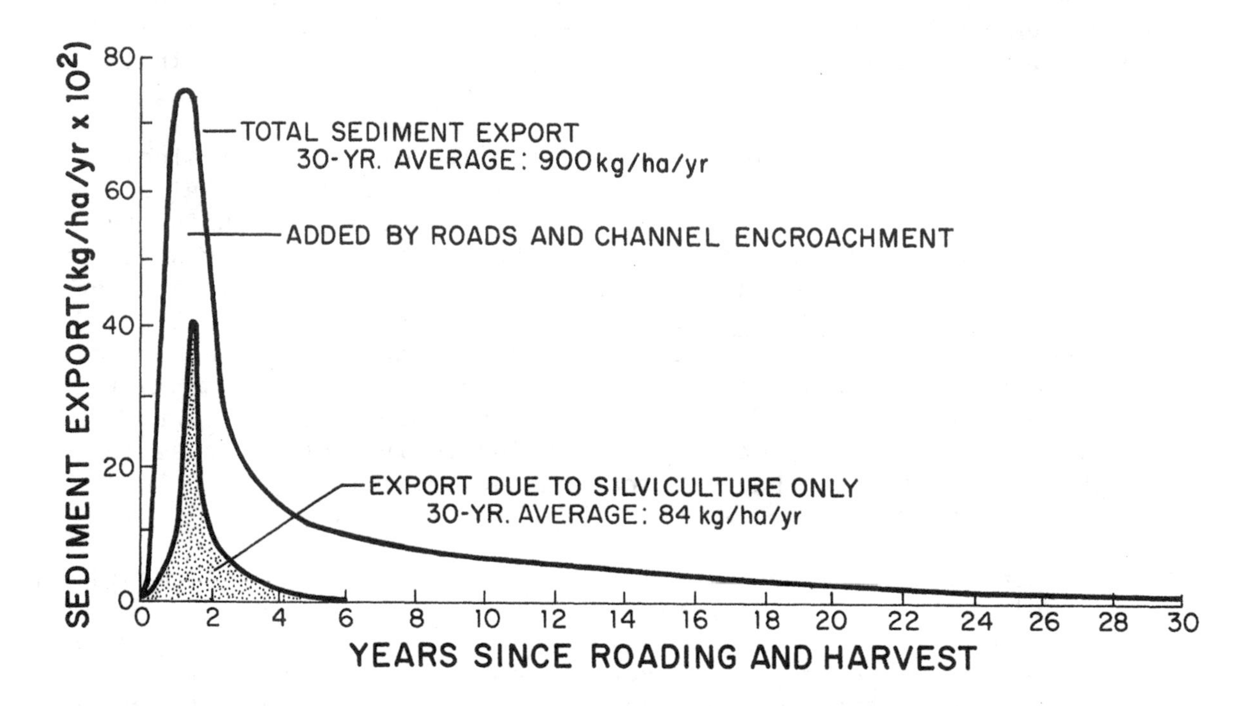

Fig 8-3. An experiment in harvesting and regenerating Piedmont forest showed that overall export was 90% due to poor roading and channel damage.

Best management practices will always include two remedies to hold sediment exports from forest operations to an acceptable level: 1) Advanced planning and careful construction of access systems, and 2) zoning of the forest property for special management of the streamside (variable source) areas.

2. The role of fire in erosion. Fire was always a natural part of the forest and grassland ecosystem. During the propaganda period of forest influences, erosion hazards due to fire were exaggerated and the illogical slogan "Fire destroys watersheds" may still be seen along highways. In some areas, notably the steep brush-covered slopes of southern California, fire may increase erosion by:

Loosening surface soil and rocks, causing shallow mass wastage.
Exposing bare soil to raindrop impact and rill erosion.
Rendering the top mineral soil layer temporarily nonwettable, thus causing overland flow during the

first rains.

Sealing the surface pores of the soil by splash erosion.

Fire-induced erosion is worst in brush and grasslands under a Mediterranean-type climate (frequent droughts and low RH). In areas humid enough for tree growth, fires must be severe and repeated to seriously accelerate erosion. The A horizon under humid forest is interlaced with coarse and fine roots which seldom burn even in hot fires. The root mat usually holds soil against erosive rains until sprouts and low vegetation reclaim the site. Regrowth is usually vigorous under conditions favorable to forest. In the 40 to 50 cm annual rainfall region, low soil moisture and poor fertility may delay regrowth and increase erosion for a time. Annual burning may reduce the annual crop of plants or litter to the point of accelerating erosion, but occasional burning or natural fires seldom lead to much erosion unless followed by abusive cultivation, overgrazing or a haphazard timber salvage.

3. The role of frost in erosion. In temperate and northern climates, freezing and thawing of exposed surface soils loosens soil and repeatedly sloughs the surface millimeter downslope. In addition, deeper frost can heave small plants out of the ground, leaving the soil bare year after year. Exposed subsoils are most susceptible to frost heaving and will often fail to stabilize without mulching, fertilizing and seeding. Cut slopes in forest roads may be sources of sediment export if the cuts are near streams. Elsewhere, small amounts of soil will settle to the base of the cut and there gradually revegetate. Frost heaving and sloughing on old gully systems may be a cause of continued erosion in some instances.

F. Summary of factors involved in erosion and sedimentation. For quick reference the following table will remind the student what to look for in assessing erosion potential under non-urban land uses.

CAUSES AND HAZARDS	CYCLIC PHASES OF EROSION
Energy sources for erosion	Detachment: Intensity and velocity of rainfall, streamflow and gully flow. Wischmeier's R, stormflow volume.
	Transportation: Volume and peak rate of stormflow; sediment load already carried by flowing water.
	Deposition: Velocity drop due to break in grade, vegetal debris and infiltration.
Susceptibility of soil	Detachment: Intrinsic erodibility of soil; the SCS erodibility factor K; exfiltration from eroding surface.

Transportation: Size of soil particles, slope of the eroding surface and forms of rills and gullies.

Deposition: Velocity drop due to vegetal debris, breaks in grade and surface roughness.

Agricultural hazards

Detachment: Annual exposure of soil by tillage; land shaping; animal and machine compaction; slope of fields.

Transportation: Sealing of soil pores causes overland flow, rills and gullies; splash erosion; slope of fields.

Deposition: Deposits at terraces, base of fields or in vegetated draws below fields. Most fines (< 25 μ) exported to still water.

Access hazards

Detachment: Width, length per unit area, grade, ditching, period of use, surfacing and stabilization of roads and trails.

Transportation: Road grade and location with respect to the variable source area.

Deposition: Deposits at grade breaks, turnouts, in vegetated draws, in saddles and below stream crossings.

Channel hazards

Detachment: Stormflow peaks and frequency; volumes, bank vegetation, debris dams, unstable trees, natural levees, old sediments; channel expansion during storms.

Transportation: Channel gradient; width, depth and velocity of flow; size of particles.

Deposition: At breaks in grade, meanders, pools, ponds, debris dams, backwaters, swamps and on natural levees.

Silvicultural hazards

Detachment: Infrequent exposure of soil by machines in harvesting, site preparation and planting; wildfire, fire lanes and prescribed burning.

Transportation: The pattern of residual debris, windrows, site preparation and planting pattern; orientation of bare soil plots; width of streamside management zones.

Deposition: Coarse material usually deposits on-site, fines more frequently off-site, determined by the distance and pattern of forest debris between point of detachment and perennial streams.

Ranked in importance as sources of sediment exports (both coarse and fine), the hazards are: Agriculture, roads, channel disturbance and silvicultural activities. Construction is probably second only to agriculture as a source of sediment, but reliable comparisons are not available.

Further readings.

Brown, G. W. Forestry and Water Quality. O.S.U. Book Stores, Inc., Ore. State Univ., Corvallis, Ore., 124 pp., 1980.

Heede, B. H. Gully development and control. U.S. Forest Service Research Paper RM-169, 42 pp., 1976.

Hewlett, J. D. Forest Water Quality: An experiment in harvesting and regenerating Piedmont forest. School of Forest Resources, Univ. of Ga., Research Paper, 22 pp., 1979.

Hewlett, J. D., W. P. Thompson and N. Brightwell. Erosion control on forest land in Georgia. Soil Conserv. Serv., U.S.D.A., Athens, Ga., 25 pp., 1979.

Holeman, J. E. The sediment yield of major rivers of the world. Water Resources Research, 4:737-747, 1968.

Leopold, L. B., M. Gordon and J. P. Miller. Fluvial Processes in Geomorphology. W. H. Freeman and Co., San Francisco, Calif., 522 pp., 1964.

Lull, H. W. and K. G. Reinhart. Forests and Floods. U.S. Forest Service Research Paper NE-226, pp. 62-81, 1972.

Wischmeier, W. H. and D. D. Smith. Predicting Rainfall-Erosion Losses from Cropland East of the Rocky Mountains. U.S. Dept. of Agric., Agric. Research Serv., Agriculture Handbook No. 282, 47 pp., 1965.

Problems.

1. A newly-built 7-ha reservoir has a capacity of 10 hectare-meters (ha-m). The forested drainage basin above the reservoir is 300 ha. Assume the entire basin is managed as described in Fig 8-3 and yields an average 900 kg/ha/yr. The sediment deposited in the reservoir has a dry bulk density of 0.9 g/cm^3. How many years will it take to fill the reservoir with sediment?

2. The Copper Basin of eastern Tennessee was denuded by SO_2 gases from copper smelting over 60 years ago. During the 17-yr period beginning in 1934, the central 1800 ha was eroded an average 0.56 m, according to data collected by the Tennessee Valley Authority. Dry bulk density of the soil in place was 1.3 g/cm^3. Calculate the sediment export in m-tons/km^2/yr, and compare to the average for N. America. What capacity (ha-m) reservoir would that export fill in one year if the sediment had a dry bulk density of 0.9 g/cm^3?

3. Large boulders are moved in mudflows partly because of bouyancy. What would a 1000-kg boulder (density = 2.6 g/cm^3) weigh if emersed in a mudflow with a wet bulk density of 1.4 g/cm^3?

4. Streamflow from a 10000-ha basin averages 50 m^3/min over the year. The basin is half agricultural and half undisturbed forest (no harvesting). Depth-integrated samples taken weekly for a year show that the average sediment concentration was 3000 milligrams per liter of water. Compute the sediment export in metric-tons/ha/yr. Was it "tolerable" for the agricultural part of the basin (Chap 8, Sec D)? Discuss the problems involved in drawing a firm conclusion from such data.

5. Select and trace a 100 to 200-ha basin from a local 7½-min USGS topo sheet. Draw in an appropriate streamside management zone. Using the principles outlined in Chaps 3, 5, 7 and 8, design a forest access system that minimizes hydrological impacts (sediment, road and channel damage) and provides about 2.5 km/km^2 of road to land (equivalent to 4 mi/mi^2). Determine approximately the average skidding distance (m) from tree stump to road.

CHAPTER 9

FORESTS AND FLOODS

A. Floods. Few words in hydrology have become loaded with so many meanings as the word "flood." Risk of human life and property damage have contributed to colorful, exaggerated and inaccurate news reports on floods. Because of the increasing encroachment of people and property on flood plains, and increased news coverage of flood events, the public regularly needs reassuring that no evidence for increasing frequency of flood flows exists. On the contrary, annual flood crests have been reduced during the 20th century by reservoirs, levees, floodways and other means. Flood damages go up every year because more and more property is built in the path of floods, and more public relief goes to communities that keep records of flood damage.

There is no doubt about the dangers of flooding; worldwide, more human life is lost in floods than in all other natural disasters combined.

There is a natural order of frequency in major floods that is related to, but not identical with, the frequency and return period of major rainstorms (Chap 4). Flood plains were overtopped in accordance with this natural frequency long before man's activities centered around the rivers; when prolonged rainfall or snowmelt exceeds the storage capacity of the basin, the excess must be passed downstream. The activities of man increases flood damage, and flood protection measures reduce it, but man has no influence over extreme rainfalls or snowmelts that cause major floods. Thus we do not "prevent" major floods; rather we devise various methods to protect life and property from them.

In the case of the more frequent local flooding on small watersheds, experiments have shown that agriculture, urban development and construction increase the frequency and magnitude of peak rates of flow. Denudation and compaction of soil are often blamed as prime factors in floods. Accounts of flood damage due to alterations in land use are exaggerated at times, chiefly because greatly increased overland flow is attributed to land use. Land-use effects on flooding are best evaluated in relation to local, on-site conditions.

1. What is a flood? The dictionary defines it as an "inundation" -- one's basement may be flooded by a burst water pipe, for example. Stormflow (direct runoff) and peakflow (Chap 7) do not have the same meaning as flood flow. Two definitions of flood in the

hydrologic sense must be distinguished: 1) Any flow of water that damages human property, and 2) any level of a natural body of water that exceeds its "normal banks."

The second is the preferred hydrological definition. Since the "normal banks" of a stream are not always clear, the U.S. Geological Survey has adopted the practice of establishing local bench marks to designate bankfull levels. Flood stages are usually reported above this flood datum. Fig 9-1 shows the frequency of recurring flood stages as related to bankfull capacity.

Flood damage does not have the same meaning as flood. Damage to human property is caused by:

Flood stage or water level, which usually occurs at maximum discharge (peakflow, Q_p).
Duration of inundation, or how long the waters remain at or above the level of the property.
Sediment delivery and deposition, which determines how much field or mud damage will occur.
Kinetic energy of flood flows, or how much force is delivered to buildings, fields, bridges and dams.
Increased earth mass which results in collapse of banks, fills and hillsides.
Failure to anticipate floods, particularly infrequent ones.
Failure to zone flood plains or to flood-proof property.

2. Types of flood. Five types of floods are recognized, each associated with its suspected cause. Some of the worst floods have occurred because of a combination of these.

Long-rain floods are associated with several days or weeks of low-intensity rainfall (cyclonic or frontal type storms). Detention storage capacity of drainage basins is finally exceeded and additional rain moves quickly into the streams. Long-rain floods are the most common cause of major flooding in most parts of the world.

Snowmelt floods are due to melting of snowpacks due to rapid increases in air temperature over snow fields. Often aggravated by accompanying rainfall, snowmelt floods are common in the northern Mississippi-Missouri basin.

Flash floods are associated with convectional rainstorms or with local bursts of rain during cyclonic storms. Because they do damage soon after and close to the causative rainfall, their sudden appearance gave rise to the ill-defined term "flash flood." Flash floods occur chiefly in summer but can occur anytime, even in the midst of long-rain floods.

Frozen-soil floods are associated with a special type of soil freezing called concrete frost, which may occur in open fields as a layer of almost impervious ice in top soil, or sometimes

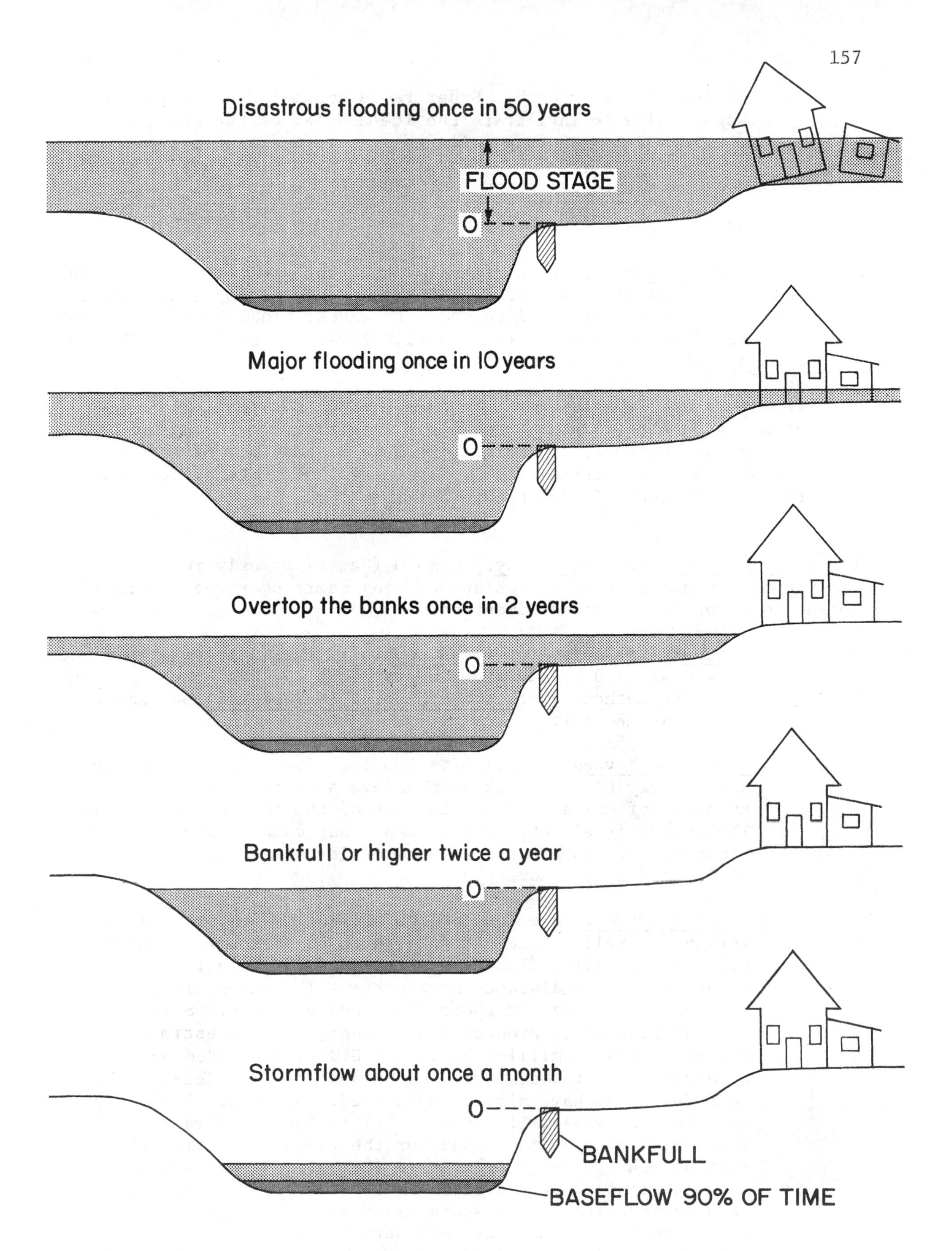

Fig 9-1. Over geologic time, channels and flood plains have evolved to carry flows of varying frequencies. In the eastern U.S., the above conditions will recur on average according to the time intervals shown.

on top of a snow pack. Under rapid snowmelt and/or rainfall, water fails to infiltrate the ice and leaves the basin as overland flow. Forest soils generally freeze into a porous granular frost, which may promote infiltration, and which will not contribute to frozen-soil floods. Frozen-soil floods are rare and are restricted to the northern-tier states, particularly the plains.

Tidal floods occur in coastal areas, and are often aggravated by upland flood waters piling against the rising tide. Associated with spring tides and hurricanes, tidal floods are common and frequently disastrous, particularly on the Gulf of Mexico and the southeastern coastline.

Foresters and land managers are concerned with floods, first because of on-site damage to soil, roads, structures, campgrounds and stream channels, and second, because of possible effects on downstream sedimentation and flood crests. Off-site effects have been the subject of much controversy.

B. The flood control controversy. Chap 8 (Sec A) briefly relates the early conflict among conservationists and engineers over the effect of forest cover on floods and sedimentation.

1. Structural measures for flood control. The major role in flood control was assumed by civil engineers about the beginning of this century. The methods which evolved for alleviating flood damages were mostly structural.

Dikes and levees to wall off flooding rivers have been used since early times. Work on the levees of the Mississippi, hundreds of miles of them, is part of the folklore of America. The method is effective, but repair and maintenance are continuous, and disastrous failures of levees and dikes have occurred fairly frequently.

Flood reservoirs are designed to store flood waters temporarily and release them at safe rates of discharge. While effective locally, flood stages are reduced only for a relatively short distance downstream. The reservoir also increases flooding for short distances upstream. Some land must be flooded to protect other land; a recent estimate indicates that 2 million ha in the U.S. are flooded by reservoirs to protect 5 million ha downstream. Because the best dam sites have already been used, the ratio of flooded to protected area will approach 1:1 in future years. Small reservoirs take up more land for the flood protection they provide than do large reservoirs (Leopold and Maddock 1954). Sediment trapped by the flood control reservoir gradually reduces its capacity to store flood waters; rates of sedimentation in the Southwest are serious enough to fill some reservoirs within 50 to 100 years. The risk of failure of dams constitutes a serious threat and the cost of spillways to prevent total failure often equals the cost of the dam itself.

Aside from their multipurpose advantages as a water supply, recreational facility, fishery and navigation channel, most of which are incompatible with flood control, reservoirs constitute very expensive protection against flood damage. An unfortunate side effect has been to lull the public into a false sense of security once a dam is constructed, which leads to further occupation of the flood plain. Consequently, although the frequency of flooding is reduced, the annual cost of flood damage continues to rise.

Assuming that floods are now "prevented" by the structure, the public risks disaster should the reservoir's capacity be exceeded by the rare but always possible extreme flood. Failure of structures due to aging, earth quake or faulty operation occurs too frequently. On average, one large dam fails each year somewhere in the world.

Floodways are alternate channels which are zoned and protected by vegetation or structures to carry exceptional flood peaks. They are dry during non-flood periods. Floodways are effective, relatively cheap, and do not preclude all use of the land during normal weather. They do, however, require maintenance and regulation.

Channel improvement (straightening, dredging or snag removal) aims at lowering local flood stages by increasing the capacity of the channel and the average velocity of the water. Straightening a channel invariably increases slope and water velocity. Temporarily effective in reducing local flooding, channel work requires frequent redredging and causes much sediment movement. In some cases, flood peaks further downstream are aggravated by more rapid drainage of channel storage above. Also channel "improvement" encourages further encroachment into flood plains.

Flood-proofing refers collectively to structures or planned utilization of flood plains to the end that occasional flooding does little damage. An example is a building on concrete pillars, leaving the ground floor free to flood. The area might be used for parking or walkways, while unmovable property is located on higher floors. Flood-proofing is not used as much as other structural means.

Structural control of floods has usually been aimed at relieving flood damage at some point along a river system, and its proponents therefore tend to think only of controlling water already in the channel system. This downstream view of flood control resulted in conflict with soil conservationists, foresters and other land managers, who naturally have an upstream view of floods.

2. Flood control by land management. The Flood Control Act of 1936 recognized the role of land management agencies in flood control. Soil conservation was enthusiastically supported during the 1930s, and the apparent connection between floods and soil erosion led to over-enthusiastic claims that soil conservation was the only sensible way to control floods. "Stop the raindrop where it falls" became the slogan of upstream flood control programs. The phrase had a plausible sound, but it was based on the erroneous idea that floods were caused exclusively by impeded infiltration and the resulting overland flow. Because land management affects infiltration rates only locally, and often only temporarily, stopping the rain where it falls sounded more effective than it proved to be. Soil conservation, which became a major activity in agricultural programs, was fully justified because of its success in reducing man-caused erosion and in rebuilding soil fertility. However, its control over flooding is limited to controlling on-site damage to fields and crops, and to reducing the sediment loads carried by streams. Soil conservation practices are mostly non-structural.

Contour cultivation favors infiltration and slows overland flow by means of ridges plowed against the flow of water. Plow furrows running upslope can quickly erode to the depth of the plow sole.

Strip cropping takes advantage both of contour plowing and alternate strips of vegetation. The vegetated strips break up the flow path and provide areas for infiltration at all seasons of the year.

Contour terracing, sometimes in addition to the above, reduces the average length of surface flow paths and the average slope of the cultivated zone between terraces, thus slowing water velocities and preventing erosion. The terraces also provide some depression storage.

Grass waterways are narrow strips of land which conform to the drainage or gully network. The strips are planted to permanent vegetation (usually a perennial grass) and left uncultivated. They serve both to catch sediments and to pass water safely out of the field.

Basin listing refers collectively to methods used to score the land surface, usually on the contour with machinery, thus to increase depression storage and infiltration. Contour ditches with intermittent cross walls to prevent water from moving along the ditch are another form.

Sub-soiling has been used in an effort to break up impermeable layers in the soil profile, thus favoring rapid percolation and better infiltration at the surface. Steel plows are pulled through compacted layers as deep as 0.5 m. While effective temporarily, the layers soon become compacted again.

Vegetal control is perhaps the most important aspect of flood damage reduction by land management. None of the above methods will be successful without planting and fertilizing to maintain their effectiveness. Forest cover is most often recommended as the ultimate in erosion control, but complete stands of grass or forbs are just as effective. Regeneration of forest is favored chiefly because such cover implies long-term protection against crop harvesting, grazing or other uses, whereas other cover types are expensive if maintained for no use other than soil conservation.

Supplementary structures are used in addition to land treatment and planting. Checkdams of various sizes and materials are used to stop gully formation and retard water flow. Outlet structures of wood, stone or concrete serve to absorb some of the energy of flowing water by dropping it in steps, dispersing the energy against a concrete apron rather than against the soil. Such structures require careful location and construction to be of any value.

Gradually a blending of the upstream and the downstream views of flood control took place. The Corps of Engineers, the main agency responsible for downstream flood control, moved further upstream to build smaller reservoirs and at the same time devoted more attention to the role of land management. The Soil Conservation Service, the agency most responsible for upstream flood control, moved downstream to build larger dams and other structures to supplement land treatment measures. Cost benefit analysis of these projects now relies more on multiple use and less on flood control benefits.

3. Other methods of flood protection. Structures and land treatment, though often justified for other reasons, are not necessarily the best form of flood protection. The following two methods have saved more lives and property than all others.

Flood warning, chiefly a responsibility of the NOAA Weather Service, has improved, and has kept several floods in the U.S. from becoming major disasters. Warning is most effective in saving lives but has also reduced property losses.

Flood plain zoning is the ultimate in flood protection but unfortunately has been slow in coming because of the continued attractiveness of the flood plain for construction. Most towns, counties and states lost the right or opportunity to zone the flood plains long before public officials became aware of the problem. Nevertheless, keeping damageable property and centers of human activity out of the potential path of floods is the best form of flood protection.

Flood damage insurance must be mentioned, not as protection but as flood compensation. No private insurance company has offered reasonable flood insurance rates, chiefly because the risks are neither random nor equal in relation to the value of the insured property. A National Flood Insurance Act was passed in 1968; most of the insurance premium for individuals is underwritten by federal subsidy.

C. Role of forest versus other types of land use. Enthusiastic conservationists and some foresters have been known to maintain that forest-covered lands simply do not produce floods. But the effect of land use and cover condition on major flooding is rather minor except insofar as the lack of good vegetative cover may set soil and sediment moving. What is the exact role of forest cover in reducing peak discharge and the volume of stormflow? A full and satisfying answer to this question cannot yet be made, but certain facts are clear.

Fig 9-2, drawn from data collected before the main dams were built, shows the movement of a flood down the Savannah River from Clayton to Clyo, Georgia. Expressed in ft^3/sec, the peak was lagged first one, then two, and finally five days, while the tributary basins piled up direct runoff into the huge discharge at Clyo. Expressed in $ft^3/sec/mi^2$, the peak flow at Clayton looks dramatic but, relative to the capacity of the channels, the discharge at Clyo was more likely to cause serious damage.

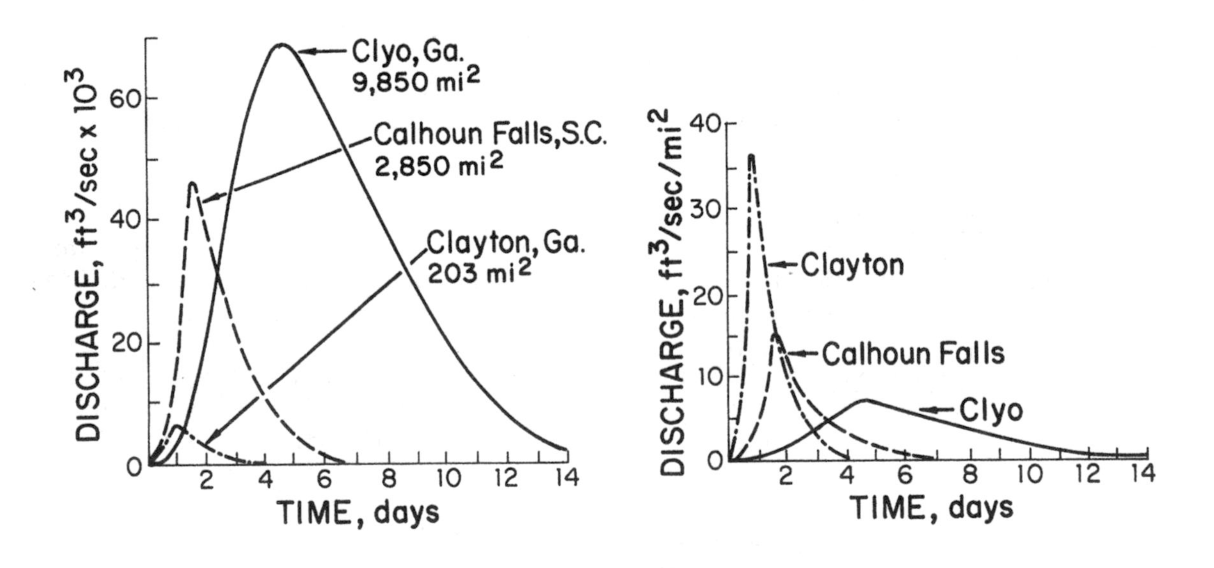

Fig 9-2. Plotted as rate per unit area (right), a flood on the Savannah River is shown as it moves over 200 miles from Clayton to Clyo, Ga. Plotted simply as rate (left), the record illustrates how peak discharges upstream were lost in the volumes of water that created the flood at Clyo.

On the main stem of a river, peak discharges do the flood damage. However, it is not the peak discharge in the headwaters that produces the downstream flood, but rather the volume of stormflow released by the headwater areas. That hydrologic fact cannot be overemphasized. As many first-order basins contribute stormflow, their respective peak discharges are staggered in time so that tributary peak rates are not additive downstream, whereas tributary volumes are additive downstream. Once the water is in the channel system, land use above obviously has no further influence on the part that particular water plays in the flood stage below. Travel time and channel storage prevent simple comparison of the hydrographs produced by first-order basins and by large watersheds.

There is no quick, simple way to make cause and effect clear, but this discussion should help to separate land influence from precipitation and channel effects on flood peaks, volumes and flood damage. Forest cover plays 4 possible roles, in order of importance.

1. Forest cover holds soil in place. The erosion that sometimes follows forest clearing and land use probably constitutes the major role of forests in floods (Chap 8). Eroding roads and channels dump sediment into lower channels where it later elevates flood stages; the settling mud and debris cause additional damage in towns, flood plains, reservoirs and in the channel itself. Roots of forest trees and shrubs may serve in some areas to bind the soil of steep slopes and help to prevent mass wastage during storms and following severe fires, but this effect is relatively minor in extreme storms.

2. Forest cover provides additional soil water storage potential. Because evapotranspiration from forest is greater than from other types of cover, the soil mantle under forest is more often dry during summer and fall. If a flood-producing rain occurs during this time, subsurface flow is diverted to retention storage under the forest. In this case, the volume of direct runoff is less under forest and the flood peaks downstream may be reduced. One watershed study showed that clearcutting a dense forest stand increased average stormflow volumes only about 10%, mostly during summer and fall (Hewlett and Helvey 1970). When the soil mantle is recharged by winter rain or spring snowmelt, forest evapotranspiration will play a minor role in reducing the volume of direct runoff. Therefore, from an engineering design viewpoint, the additional storage provided under forest cover cannot be depended on, because the largest flood flows usually occur when antecedent soil moisture is high, with or without forest cover.

3. Forest cover maintains infiltration. Abuse of the land after forest removal by repeated burning, unwise agriculture, overgrazing or haphazard road building, reduces infiltration rates and increases local peak flows and volumes. In other words, infiltration is impaired by the practices which often follow forest clearing, not merely by the cutting and removal of trees and shrubs. So far there is no evidence that the mere presence or absence of forest cover affects infiltration to such an extent that either the prevention or cause of major floods can be related directly to it

(Lull and Reinhart 1972). In order to increase downstream floods seriously, the reduction in infiltration must be drastic and must simultaneously cover large areas, conditions that fortunately rarely prevail.

4. Forest cover favors granular soil frost (as opposed to concrete frost) in the northern latitudes. Evidence from actual measurements of stormflows on basins exhibiting soil frost is scarce, therefore it is currently difficult to evaluate this particular forest influence on flooding and flood damage. Lull and Reinhart (1972) review the sparse information.

Swamplands, whether forested or not, produce surprisingly rapid stormflow because rainfall soon brings the water table to the surface all over the basin. Once at the surface, water will move rapidly into channels and downstream, regardless of the low channel slope. Steep land with shallow soil (less than 1 m) under high rainfall also produces rapid stormflow whether forested or not. Above 1400 m elevation in the southern Appalachian Mountains, it is not unusual for forest-covered slopes to average 38 cm of stormflow per year (the average for the East is about 13 cm). Despite high rainfalls that favor stormflow from some forest areas, there is no doubt that forest cover preserves natural water quality and moderates both on-site and downstream flood damage by reducing erosion and sedimentation.

In conclusion, land use has less effect on stormflows and flooding than does soil depth, texture and layering; these mantle properties vary by physiographic province (Fig 7-2). Flood damage, on the other hand, is directly affected by land use of the flood plains and by on-site structural and cultural practices that are improperly designed or located (roads, culverts, terraces, small dams, etc.). The natural stormflow response of land is difficult to alter. For example, severe surface mining that exposed up to 60% of the soil in several small watersheds in eastern Kentucky caused neither change in average storm response nor any increase in the annual peak discharge (Bryan and Hewlett 1981). Soil mantle storage was actually increased by the blasting and grinding of rock materials moved to get at the coal. While sediment export shot up dramatically, the volumes of water released during large storms was no greater than before coal mining. Studies of some urban basins (Wallace 1971) have shown similarly that while late summer stormflows in Atlanta were doubled by urbanization, the largest annual peak discharges, usually in late winter, were not seriously affected. Forest operations do not increase flood flows appreciably but may add to flood damage if roads are badly designed and if channel zones are not respected.

Further readings.

Bryan, B. A. and J. D. Hewlett. Effect of surface mining on stormflow and peakflow from six small basins in eastern Kentucky. Water Resources Bull., 17(2):240-249, 1981.

Hewlett, J. D. and J. D. Helvey. Effects of forest clear-felling on the storm hydrograph. Water Resources Research, 6(3):768-782, 1970.

Lull, H. W. and K. G. Reinhart. Forests and Floods. U.S. Forest Service Research Paper NE-226, 94 pp., 1972.

Wallace, J. R. The effects of land use change on the hydrology of an urban watershed. Environ. Resources Center, Georgia Tech, Atlanta, Ga., Report ERC-0871, 66 pp., Oct., 1971.

Problems.

1. A meandering channel drops 1 m in 1 km, measured along the thalweg (Chap 3). The 1-km reach is dredged and shortened to 0.5 km, retaining the same channel cross-section but reducing Manning's n from 0.05 to 0.02. Use Manning's formula (Eq 7-3) to estimate the percentage increase in a flood flow Q that can be accommodated if A and W_p are the same after dredging. Explain the increase.

2. The 10-yr return period rainstorm at any point in southeastern U.S. is about 20 cm in 2 days, but the stormflow produced by that rain on a small watershed may not be the 10-yr stormflow. Explain.

3. A report from a small watershed experiment concludes, "Flood peaks were increased 300% by clearcutting the forest on the basin." Assuming that the percentage was correctly calculated from data, evaluate and explain the conclusion in terms of flooding and flood damage.

CHAPTER 10

FORESTS AND WATER QUALITY

A. Introduction. Water quality is customarily defined in relation to intended use, e.g., drinking, recreation, irrigation, power generation or manufacturing. Water quality in general includes the physical, chemical and biological properties associated with the mineral and organic material suspended and dissolved in water, including organisms. High quality water for a given use may not require limits on all three properties. For instance, water used in irrigation should not have a high salt concentration, that used for cooling should have a low temperature and that used for recreation must have a low infectious organism count. Water acceptable for one use may not be for another.

Water quality is influenced by climate, season, soil and rock mineralogy, vegetation and the activities of man. When natural water is fouled by man's activities to the point where it can no longer meet a specific use, it is said to be polluted. Pollution, like water quality, is a term related to the intended use.

It has long been recognized that forests promote high quality water, primarily by reducing erosion and sedimentation. Burning, harvesting or clearing forests may temporarily degrade water, but it is the practices which follow forest removal that produce severe and lasting degradation.

The subject of water quality and how to measure, predict and control it is too wide to include in an outline of hydrology. This chapter is added to introduce terms and to direct the student's attention to a companion outline, Forestry and Water Quality, by George W. Brown (O.S.U. Book Stores, Inc., School of Forestry, Oregon State University, Corvallis, Oregon, 124 pp., 1980). Brown's text covers erosion, sedimentation, water temperature, nutrient cycling, dissolved oxygen and silvicultural chemicals in relation to water quality.

B. Measures of water quality are related to intended use; no single measure describes overall water quality. Physical, chemical and biological analysis of water is a rapidly expanding technology. Manuals for water analysis are issued by a number of agencies; but the U.S. Environmental Protection Agency is the primary source of information on methods and standards (Envir. Prot. Agency 1976). There are as many methods as there are materials in water.

1. Water characteristics:

Organic suspended sediments consist of particulate vegetal and faunal matter, usually carried in greatest concentrations during early stormflow (before peak discharge).

Inorganic suspended sediments (Chap 8) consist of sands, silts and colloids of various minerals which in time will settle out of water. These too are carried in greatest concentration during early stormflow.

Turbidity is the degree of opaqueness, or cloudiness, produced in water by suspended particulate matter, either mineral or organic. Turbidity is measured by light transmission through a sample of the water. When calibrated against filtered samples, turbidity may be expressed in milligrams per liter (mg/l) or parts per million (ppm). At high turbidities, calibration is difficult and unsatisfactory, limiting the usefulness of turbidity as an index of total load.

Total dissolved solids is a measure of the mineral and organic material dissolved in water. It is absolutely determined by evaporating a filtered volume of water and weighing the residue. The amount of residue is expressed as a proportion of the original sample in ppm or mg/l.

Dissolved gases, including oxygen, carbon dioxide and nitrogen, are each measured by special methods (Envir. Prot. Agency 1976). Portable devices are available for oxygen measurement.

Alkalinity is the ability of solutes to neutralize acids, or the ability of water to resist a shift in pH. Alkalinity usually reflects the activity of calcium carbonate.

Hardness is caused primarily by the concentration of calcium (Ca) and magnesium (Mg) ions. Soft water has a concentration less than 60 mg/l, whereas hard water has a concentration greater than 180 mg/l. Soft water is desirable for most uses; on the other hand, streams and ponds with a hardness of 20 mg/l or less are often biologically unproductive.

Total conductivity (in µmhos/cm at 25°C) is a measure of the electrolytes in water, and is sometimes assumed to indicate total dissolved solids. However, minerals and organic solutes vary in electrolytic activity, so that exact calibration of conductivity is possible only for one solute species at a time. Total conductivity will increase with total dissolved solids, alkalinity and hardness, and decrease with increased temperature of the solution.

Immiscible liquids, such as oils, fats and various organics, degrade the quality of water by forming scums, interfering with gaseous exchange with the air and in severe cases destroying birds, fish and other wildlife.

Bacteriological water quality constitutes a highly technical field related to bacteria, viruses and fungi in water, particularly as these affect man, fish and wildlife. For example, well and spring water is often tested for "fecal coliforms," chiefly of the genus Escherichia spp, because health officials believe these to be good indicators of viral and other disease vectors. Nearly all natural water contains some fecal coliforms, but those with certain types, or higher than normal levels, are considered hazardous to human health. Typhoid fever is an example of a water-born bacteriological disease. Because even insects discharge fecal coliforms, virtually all open channel water is condemned for drinking (without boiling) by public health offices. Campsite water supplies are therefore expensive to install and maintain.

Dissolved oxygen (D.O.) is an indicator of the suitability of water as a habitat for higher aquatic organisms, particularly insects, crustaceans and fish. Normal concentrations of dissolved oxygen, between 8 and 14 ppm, indicate that aerobic conditions exist. Concentrations less than 5 ppm are toxic to most fish. Zero dissolved oxygen marks the boundary between aerobic and anaerobic (septic) conditions. (Forest Service, Chap IX, 1980.)

Biological oxygen demand (B.O.D.) is mostly used to refer to pollution and waste water treatment. It is a measure of the amount of organic matter to be oxidized through aerobic biochemical processes. In general, a high B.O.D. indicates a high concentration of organic matter in the water. Natural water has a B.O.D. of about 2 to 3 ppm, while untreated sewage and pulp mill effluents often exceed 200 ppm. It is common for treated sewage to be described as having a certain "percentage B.O.D. removed." For instance, 95 percent B.O.D. removal means that only 5 percent of the original raw sewage B.O.D. remains.

Toxic chemicals, particularly the long-lived varieties of man-made pesticides, detergents, hydrocarbons, heavy metals and highly stable, non-biodegradable materials used in the plastics industry (PCBs, etc.), are found in small amounts all over the world in water, soil, air, plant and animal tissues. Analysis and assessment of the dangers of these chemicals is a highly technical field beyond the scope of forest hydrology. Fortunately, only a few phytocides and insecticides are now thought useful in forest and wildland management. However, specific guidelines for the use of these few, particularly in reference to potential water pollution, require knowledge of hydrologic process to be effectively applied (Forest Service, Chap XI, 1980).

Temperature has an effect on all aspects of water quality, particularly those influenced by biological activity. The concentration of dissolved gases generally increases with decreasing temperature.

The pH of water is frequently reported as an index to pollution by acid- or base-forming chemicals, chiefly sulphur and nitrous oxides in the acid range, and Ca and Mg in the basic range. Units (0 to 14) are negative logarithms of the hydrogen ion concentration in water;

pH 7 divides basic from acid reaction. However, pH 5.6 is taken as the dividing line between "acid rain" and normal rain, since that is the reaction of pure water acidified by atmospheric carbon dioxide. Rain in the eastern U.S. falls in the "acid rain" category, generally between pH 4 and 5. It is not yet clear whether these acidity levels represent pollution by man or contamination by natural processes. Stream water has a normal reaction around pH 6, except in streams draining basic-rock terranes (pH 7 to 8). It is important to know the solutes causing undesirable effects and how well buffered the solution is at a given pH. For example, a weak solution of calcium carbonate will have a pH of 8, but so will a strong solution. For this reason, pH by itself is not an absolute measure of pollution or contamination by natural solutes.

2. Mass, volumes and units used in water quality work:

Milligrams per liter (mg/l) and parts per million (ppm) are not exactly the same measure but at low concentrations normally found in natural waters they have nearly the same numerical value.

A mg/l means 1 milligram of suspended or dissolved material in a total volume of 1 liter. Thus mg/l has units ML^{-3}.

A ppm means 1 unit of suspended or dissolved material in a total of 1 million units. Thus ppm is dimensionless, that is, M/M or L^3/L^3. The latter (volume/volume) is seldom used.

> Example: The difference between mg/l and ppm (M/M) can be shown by adding 1 g of NaCl to 1 liter of pure water. The dissolved salt adds virtually nothing to the volume of the mixture, but does add to the mass. Calculate mg/l and ppm, remembering that 1 g of H_2O = 1 cm^3.

$$\text{mg/l} = \frac{\text{1000 mg of NaCl}}{\text{1 liter of mixture}} = \text{1000 mg/l}$$

$$\text{ppm} = \frac{\text{1 g of NaCl}}{\text{1001 g mixture}} \times 10^6 = \text{999.001 ppm}$$

> If 1 g of silicon dioxide sand is suspended in 1 liter of water, ppm will be the same as above, but mg/l will change because the volume of the mixture will be slightly greater than 1000 cm^3 (1000 + 1/2.5 = 1000.4 cm^3, in fact, because sand has a specific gravity of about 2.5). So,

$$\text{mg/l} = \frac{\text{1000 mg of } SiO_2}{\text{1.0004 liter of mixture}} = \text{999.60 mg/l}$$

Toxic chemicals in water are often measured in parts per billion (ppb) or micrograms per liter (μg/l) because concentrations are very low. At that level the difference between the two expressions is numerically negligible.

C. Natural water quality. The streams and lakes of North America had a natural quality before any disturbance by man. Dust, atmospheric minerals and gases are picked up as precipitation falls through the atmosphere. Between rains, dry fallout occurs when dusts and aerosols are blown about. Among the most common are carbon, sulphur, sodium, calcium, oxygen, nitrogen and silica but virtually every element will at sometime be found in rainwater. During the interception process, throughfall and stemflow pick up additional minerals and organics from vegetation. As the water, already containing many elements, moves through the weathered mantle, further solution occurs and complex exchanges between the mineral, floral and faunal components of the soil take place. When the water finally emerges as streamflow, its organic and inorganic constituents reflect the mineralogy of the basin, the historical character of the precipitation and the nature of the vegetal cover, roughly in that order. Further changes take place in the channel, where riparian vegetation and the wind add organic matter during leaf fall. Abrasion and solution of sediments in the channel add more inorganics to the dissolved and suspended components of streamflow. The quality of subsurface flow emerging into the channel is apt to be fairly stable from year to year. However, suspended sediments in streamflow will vary greatly with the discharge rate.

1. Suspended material in natural waters is divided into two types:

Inorganic sediments, consisting chiefly of sands, silts and colloids from the surface of the drainage basin and from channel bottoms (Chap 8).

Organic sediments, consisting of living or dead particles of plants (both terrestrial and aquatic) and animals (insects, crustaceans, amphibians and fish) carried by the water. Organic sediments may be decomposed or ingested by the stream biota (insects and other faunal orders, bacteria, fungi and algae) to produce further particulate matter or organic solutes.

As shown in Chap 8, the greater the velocity of flowing water, the greater the amount of material which can be held in suspension. The sediment load of stream water during stormflow changes greatly even under natural conditions. It is not uncommon for three quarters of the total annual suspended material yielded by a basin to be discharged during one storm. This greatly complicates experimentation, modeling and prediction of sediment export. In general, deciduous forests yield more organic sediments than coniferous forests. Temperate-zone streams draining 1 km^2 of deciduous forest will receive from 5 to 10 metric tons of dry-weight organic matter per year; much of that leaves the basin as fine particulates and dissolved organics.

One effect of suspended sediment is to inhibit the growth of aquatic organisms because solar radiation, which is essential for photosynthesis, cannot sufficiently penetrate the dirty water.

2. Dissolved materials in natural waters may also be divided into two types:

Inorganic solutes include both minerals and gases. Although minerals account for the greatest percentage of the dissolved inorganics, the gases, particularly oxygen and carbon dioxide, play a key role in sustaining the aquatic ecosystem and in determining the quality of water. Carbon dioxide and water supply the major components of cell protoplasm. Oxygen and carbon dioxide regulate the metabolic processes of aquatic life.

Virtually all of the naturally occurring elements of the earth's mantle, including radioactive material, may be found in surface waters. Many occur only in trace concentrations, which nevertheless are measured in prospecting for gold, silver, lead, copper, zinc and uranium. As a percentage of total dissolved solids, the most common ions and compounds in the surface waters of North America are:

Ion or compound	Average % of total
CO_3^{--}	33.4
Ca^{++}	19.4
SO_4^{--}	15.3
SiO_2	8.6
Na^{+}	7.5
Cl^{-}	7.4
Mg^{++}	4.9
K^{+}	1.8
NO_3^{-}	1.2
Fe_2O_3, Al_2O_3	0.5
PO_4	<0.01
Total solute	100 %

The anions (negative charge) occur in combination with the cations (positive charge) to form salts. Carbonates are the most abundant salts because of the availability and solubility of carbon dioxide. Phosphate (PO_4) is mineralized or metabolized so readily that little of it appears in streams not polluted by industrial wastes. Concentrations vary considerably from season to season and basin to basin. As an example how water quality for a particular use may be affected by dissolved minerals, a small amount of calcium gives a good taste to water, but an equal amount of magnesium is distasteful to most people.

Organic solutes constitute an extremely complex series of compounds produced by photosynthesis, metabolism, and the decomposition of plant and animal tissues. Some are fleeting and unstable, some are absorbed by organisms to produce further organic sediments, some are toxic, and many organic compounds pass on to serve as nutrient and energy sources for communities of plants and animals downstream. Recent emphasis on water pollution has stimulated much research on this aspect of natural water quality.

Many natural waters do not support large and diverse populations of aquatic organisms because the inorganic and/or the organic nutrient supply is low. For example, the streams of the southern Appalachian Mountains are notably infertile because heavy annual rainfall has long ago flushed out the easily soluble minerals, diluting inorganic solutes to a level incapable of supporting stable populations of fish and other aquatic flora and fauna. On the other hand, streams draining limestone terranes of the Valley and Ridge Province of the Appalachians are noted for aquatic fertility and diverse speciation, because rainfall is less and basin mineralogy is dominated by calcium and other essential elements associated with a neutral to basic pH.

D. Polluted waters. Because natural waters already carry materials which can degrade water for certain uses, there are semantical difficulties with the word "pollution." Natural water quality over the centuries has evolved stream ecosystems under conditions we might, rather pointlessly, refer to as "natural pollution." It seems best to regard pollution as man-caused and to think of polluted waters as those degraded below the natural level by some activity of man. In this sense, unabused forests and wildlands do not produce polluted waters, although they may produce contaminated water, i.e., naturally impaired quality.

As an example of pollution, natural waters usually contain little phosphorus, a nutrient essential for plant growth. Competition among species for phosphorus in a natural stream is severe, and the ecosystem is subtly controlled by its relative availability. Several activities of man (for example, the application of commercial fertilizers to ponds and riparian areas, or the dumping of industrial wastes and sewage) have increased the supply of available phosphorus sufficiently to cause sudden lush growths ("blooms") of algae and other objectionable plants in streams and lakes. Although a valuable nutrient, these additional supplies of phosphorus constitute pollution. The death and decay of the blooms remove oxygen from the water by aerobic decomposition and fish kills may occur. The problem is self-correcting if the pollution is stopped.

In the case of radioactive materials, heavy metals and man-made chemicals, pollution may be virtually permanent, and is therefore a much more serious problem than the undesirable nutrient enrichment (eutrophication) of lakes and streams. New chemicals are being fabricated faster than old ones can be evaluated for their long-term effects. Some are banned by federal law for certain uses but approved for others, some are removed from the market temporarily or permanently, and others are sold freely even though under suspicion as carcinogens. Information on pollution by man-made chemicals becomes quickly out of date, therefore the student is directed to the latest journals on environmental pollution for assessment of current problems. Fortunately few chemicals really needed in forestry have been designated severe hazards.

1. Temperature pollution by forest activities is sometimes blamed for death of fish, particularly trout and salmon, but the full importance of the effect on fresh water fisheries is yet to be assessed. Fig 10-1 shows the measured effect of a clearcutting operation on water temperature at the mouth of first order streams in the Georgia Piedmont. The control, or uncut, basin (about 40 ha) yielded water that was warmer than both the air and the cutover basin in January. In July, the cutover basin's water was cooler than the air at noon but warmer at midnight; both were warmer than the water flowing from the control basin throughout the day. Temperatures above 80 degrees Fahrenheit (26.8°C) are considered critical or lethal for some species of aquatic life, but in Piedmont waters there appeared no damaging effects. Summer increases in water temperatures persisted for 6 years following clearcutting; a high, complete forest canopy is apparently necessary to maintain stream temperatures characteristic of forest cover.

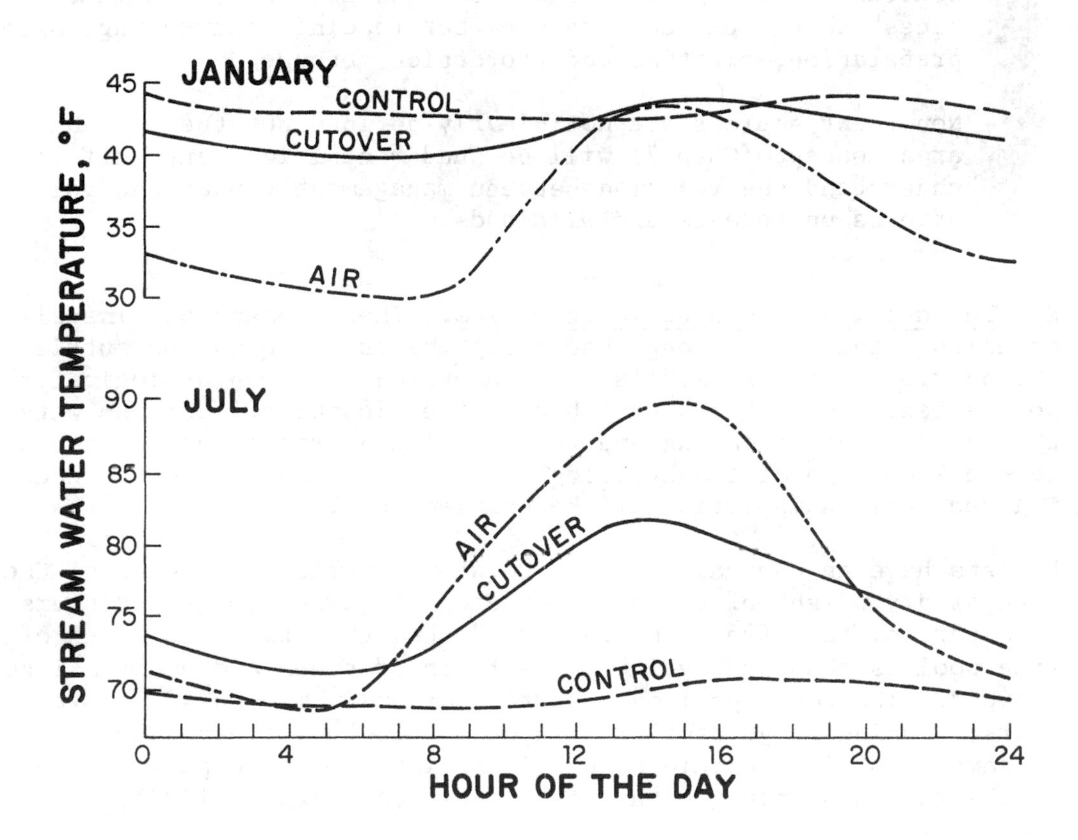

Fig 10-1. Daily temperature regimes following clearcutting in the Georgia Piedmont. The control stream was shaded by pine and hardwood; the "cutover" stream was partially exposed to direct sun (Hewlett 1979).

2. Point and non-point source pollution are legal terms that evolved from court decisions aimed at implementing the provisions of the National Environmental Protection Act (NEPA) of 1972(77),

particularly Section 404 on point sources and Section 208 on non-point sources. A point source was defined as a pollutant carried to natural waters through a "discrete conveyance," implying a pipe, channel outfall or dumping point for industrial, municipal and agricultural wastes. "Non-point source" was defined simply by exclusion as any other source for material found in natural water courses. The distinction was adopted for convenience in regulation: Point sources were to be stopped by internal alterations in plant procedures or processes, whereas non-point sources were to be mitigated by suggested or required improvements in land management practices. Called "best management practices," these non-point source remedies were to be worked out over a period of years. Since both natural contamination and man-caused pollution stem from non-point sources, and since real causes and effects are not yet well understood, confusion resulted from the loose distinction between point and non-point sources. Finally all silvicultural activities were classified legally as non-point sources to be regulated under Section 208 "best management practices" that would encourage better roading, harvesting, site preparation, planting and protection methods.

Non-point sources are not clearly defined but the variable source area concept (Chap 7) will be fundamental in future efforts to understand the relation between management causes and pollution effects on forests and wildlands.

E. The nutrient and hydrologic cycles. The movement of minerals and organics within and through the biosphere is known as the nutrient cycle. The nutrient cycle is similar in conception to the hydrologic cycle; both extend above, below and through the biosphere. Because water is the chief agent releasing and transporting nutrients around the ecosystem, a knowledge of the hydrologic cycle is essential to the understanding and safe manipulation of the nutrient cycle.

Forests have an unusual capacity to store nutrients because of the large mass of dry weight of the standing crop. Leaves, twigs and stems become part of the "free nutrient pool" when they fall, but much of this free pool is absorbed again by the roots of trees and by faunal populations. The total pool on a particular basin may build up over the years, leading to greater growth potential, but it may require a fire or severe cutting to release nutrients into the free pool for ready re-absorption by new growth (Forest Service, Chap X, 1980).

Nutrient cycling on the watershed is sometimes depicted as taking place in one dimension, from sky to plant, from plant to soil and from soil back to plant again -- a cycle in situ. Consideration of the variable source area for streamflow (Chap 7) and soil water movement in slopes (Chap 5) suggests that a spiraling cycle is a more accurate description of mineral movement. The release of minerals and water to streams is the net escape from the drainage basin ecosystem. Nutrients not delivered directly to the flowing stream have more than one chance of being captured by plants, animals and soils on the basin.

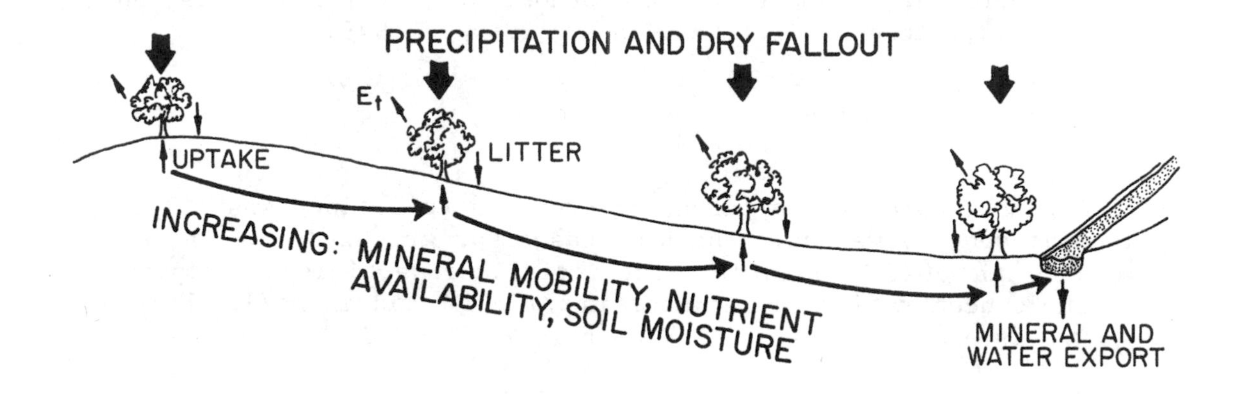

The diagram above shows how minerals delivered by rain or "dry fallout" from the air, as well as those released by weathering of the soil mantle, are shifted toward the stream channel, controlled by the pathway of water and the migration of litter downslope. The familiar improvement in site productivity (site index) downslope is one consequence; the attenuation or disappearance of pollutants applied upslope is another.

The drainage basin is the most logical unit of the ecosystem. Accordingly there is a growing tendency to combine objectives of land, air and water management into one system for identifying and solving land management problems. The forest and wildland property lends itself easily to patterned management, with careful attention paid to stream networks, water divides, topographic position and hydrologic hazards. Control of water quality and prevention of non-point source pollution by forest operations depends on land management pattern.

Further readings.

Brown, G. W. Forestry and Water Quality. Oregon State Univ. Book Stores, Inc., Corvallis, Ore., 124 pp., 1980.

Environ. Prot. Agency. Quality Criteria for Water. Environ. Prot. Agency, Wash., D.C. Printed by: Nation. Tech. Inform. Service, U.S. Dept. of Commerce, PB-263942, 501 pp., July, 1976.

Forest Service. Water Resources Evaluation of Non-point Silvicultural Sources. U.S. Dept. of Agriculture, Procedural Handbook, Printed by: Nation. Tech. Inform. Service, U.S. Dept. of Commerce, Wash., D.C., 780 pp., 1980.

Hewlett, J. D. Water Quality: An experiment in harvesting and regenerating Piedmont forest. School of Forest Resources, Univ. of Ga., Athens, Ga., Res. Paper, 22 pp., 1979.

Problems.

1. A grab sample from a turbulent creek in full flood was 10% suspended fine sand by weight. The sand has a sp. gr. of 2.60. The sand does not absorb water into the individual particles. Compute the concentration of sediment in ppm (M/M) and in mg/l. Which measure is easier to use in sediment work?

2. Secure a copy of your State's Section 208 best management practice guidelines. In a short field trip, examine a recent forest management operation and report on the adequacy of compliance with the guidelines.

TABLE OF APPROXIMATE CONVERSION FACTORS

To convert column 1 into column 2, multiply by	Column 1	Column 2	To convert column 2 into column 1, multiply by
	LENGTH		
0.621	kilometer	mile	1.609
3.281	meter	foot	0.3048
0.394	centimeter	inch	2.54
	AREA		
10.764	$meter^2$	$foot^2$	0.0929
2.471	hectare	acre	0.405
0.386	$kilometer^2$	$mile^2$	2.590
	VOLUME		
0.0610	$centimeter^3$	$inch^3$	16.387
35.314	$meter^3$	$foot^3$	0.0283
0.2642	liter	U.S. gallon	3.785
8.107	hectare-meter	acre-foot	0.12335
0.2400	$kilometer^3$	$mile^3$	4.1655
	WEIGHT*		
1.1020	metric ton	U.S. ton	0.9074
2.2045	kilogram	pound	0.4536
0.0353	gram	ounce (avdp)	28.35
	PRESSURE		
14.50	bar	$pound/inch^2$	0.06895
0.9869	bar	atmosphere (std.)	1.013
	RATE		
0.8921	kilogram/hectare	pound/acre	1.121
0.4464	metric ton/hectare	U.S. ton/acre	2.240
0.5886	$meter^3/minute$	$foot^3/second$	1.699
1.524	$meter^3/minute/km^2$	$foot^3/second/mi^2$	0.6562
62×10^{-6}	$gram/meter^3$ (or mg/l)	$pound/foot^3$	16×10^3
	TEMPERATURE		
(9/5) C + 32	Celsius	Fahrenheit	5/9(F - 32)
	DENSITY		
62.42	$gram/centimeter^3$	$slug/foot^3$	0.0160

*Metric ton, kilogram and gram are both mass and weight units at sea level; U.S. ton, pound and ounce are weight units only. At sea level, a pound weight is one slug.

INDEX

CPSIA information can be obtained
at www.ICGtesting.com
Printed in the USA
LVHW052258310821
696580LV00009B/775